S0-AYS-868

Alexander Waugh was born in 1963, and is the grandson of Evelyn Waugh and the son of the columnist Auberon Waugh and novelist Teresa Waugh. After reading Music at Manchester University, he became the Chief Opera Critic at the *Mail on Sunday* and then the London *Evening Standard*. He has written several books on music, including *Classical Music: A New Way of Listening*, published in 1995, which has been translated into fourteen languages. A companion volume on opera was published in 1997.

Alexander Waugh is also a publisher, a cartoonist and an illustrator, and composed the music for the award-winning stage musical, *Bon Voyage!* He lives in Northamptonshire with his wife and three children.

TIME

FROM MICRO-SECONDS TO MILLENNIA –
A SEARCH FOR THE RIGHT TIME

ALEXANDER WAUGH

HEADLINE

Copyright © 1999 Alexander Waugh

The right of Alexander Waugh to be identified as the Author
of the Work has been asserted by him in accordance with the
Copyright, Designs and Patents Act 1988.

First published in 1999
by HEADLINE BOOK PUBLISHING

First published in paperback in 2000
by HEADLINE BOOK PUBLISHING

10 9 8 7 6

All rights reserved. No part of this publication may be
reproduced, stored in a retrieval system, or transmitted,
in any form or by any means without the prior written
permission of the publisher, nor be otherwise circulated
in any form of binding or cover other than that in which
it is published and without a similar condition being
imposed on the subsequent purchaser.

ISBN 0 7472 5988 7

Typeset by Palimpsest Book Production Limited,
Polmont, Stirlingshire
Printed and bound in Great Britain by
Clays Ltd, St Ives PLC, Bungay, Suffolk

HEADLINE BOOK PUBLISHING
A division of Hodder Headline
338 Euston Road
London NW1 3BH

www.headline.co.uk
www.hodderheadline.com

For Eliza

CONTENTS

CHAPTER I

INITIUM
Beginning

'In the beginning God created Heaven and Earth, and the Earth was without form and void; and darkness was upon the face of the deep.'

Yes, yes, we all know that, but what happened *before* the void, *before* the darkness was upon the face of the deep? Or, as the philosopher Bertrand Russell asked when he was a seven-year-old schoolboy, 'If God made everything, who made God?'

The Bible does not tell us what happened before the beginning, which is deeply frustrating, for we would all love to know. According to the myth of Genesis there *was* no before, or, at least, not in any sense that we could understand. We find it hard to grasp the idea that time, once upon a time, may have had a beginning and that one day it will probably come to an end. 'That's not possible,' the inquisitive mind will urge; 'if time had a beginning, what was happening two hours *before* it began? How can it have started without something having happened to make it start in the first place? And how could that something not have existed before the start which it was supposed to have created?'

It is not difficult for an averagely inquisitive child to find himself embroiled in a confusion like this after only a few minutes of stargazing. Gawping up at the sky on a balmy night during the summer holidays, children are particularly adept at latching on to the antinomies of time and space and asking questions that their parents, however well meaning, cannot begin to answer. For the parent, who has long ceased to ponder these problems and is living in denial, a child's fertile questioning can lead to irritability:

'Dad, what's at the end of the universe?'

'Er, more stars and stuff like that.'

'No, I mean after all the stars, what is beyond them?'

'Oh, sorry . . . um . . . space.'

'And what's beyond the space?'

'Nothing, just more space. Go to bed.'

'Yes, I know, but what's at the end of that?'

'Listen, God made it all. He is infinite, and he made the universe, which is also infinite, OK?'

'How did he make it?'

'He just did – he is all-powerful.'

'But how big is the universe?'

'I told you. It is infinite. You cannot measure it – it goes on for ever.'

'Can God measure it?'

'No.'

'But you said he is all-powerful.'

'Yes, then . . . but . . .'

'So, if God is all-powerful, can he make another universe that is so complicated that he cannot measure it?'

'Yes, no – yes – he is all-powerful.'

'You said he is all-powerful, but if he cannot make such a universe, then he cannot be all-powerful, can he? And yet, if he *is* able to make a universe that is so complex that he cannot measure it, then he is still not all-powerful because he cannot measure it. So, either way he cannot be all-powerful, and I thought you said . . .'

'I thought I told you to go to bed.'

Children are remarkably perspicacious when it comes to getting under the skin of these sorts of problems. Adults, on the other hand, are adept at brushing them aside. 'They fuck you up, your mum and dad,' as the poet Philip Larkin so famously put it, and one of the ways they most persistently 'fuck you up' is by not concentrating when you ask them interesting questions. 'Go to bed' cannot be a satisfactory answer to the child's paradox. In the back of our minds we are all aware that questions about infinity, simple questions about the measurement of time and space or God's omnipotence, have not been satisfactorily resolved. We want to understand them, we would love to be able to stuff infinity into a nutshell and squint at it through a lens, or, as William Blake put it,

> To see a World in a Grain of Sand,
> And a Heaven in a Wild Flower,
> Hold Infinity in the palm of your hand,
> And Eternity in an hour.

But a pragmatic adult usually thinks to himself, 'I can either spend the rest of my life going round and round in circles trying to work it out or – forget it.' But this is a form of mental laziness. In point of fact the child's argument can be taken apart without too much difficulty, but it needs the sort of faint effort that parents are not always willing to give. I offer three possible ripostes, each of them better than 'Go to bed':

1. If Dad had been a very religious man he could have answered like this: 'God is all-powerful – that is an irrefutable tenet of my faith. The fact that you are unable to understand his omnipotence or the limitless space of his universe is simply because your own little brain is finite – it occupies a finite space and is filled with a finite number of brain cells. Obviously, you cannot expect to understand infinity with your little finite brain. Only God can understand infinity because his own wisdom is alone infinite. It is not for us to understand infinity – we will never be able to do that – but only for us to believe in it as an act of faith in God.'

If that does not succeed in sending the wretched youth to bed, perhaps this one will:

2. 'My child, you fail to understand the nature of power. Power is about possibility. *Puissance*, the French word for power, is derived from the root word *pouvoir* – "to be able". The more power you have, the more you are *able* to make *possible* things happen; if you are all-powerful, as God is, you are able to make all *possible* things happen, but no amount of power can make the *impossible* happen. For instance, God, who is all-powerful, cannot draw a circle which is shaped like a square; that is clearly impossible and therefore beyond the realms of power to achieve. You are suggesting that God is not all-powerful only because he cannot do something that is impossible. Your argument proves to me only that you do not understand the meaning of the word power.'

This, especially if delivered in a monotone, should put the child off his keen-nosed questioning. It will certainly make him think twice next time. But there is an alternative, perhaps more compelling argument that a father might employ if he wishes to finish the thing off for good:

3. 'Young whippersnapper, if God is all-powerful he must possess the power to create a universe which is so large that it cannot be measured – not even by himself. When, and if, he creates such a universe he will no longer be all-powerful, since, in that event, he will not possess the power to measure his new universe. But that is only a possibility for the future; it relies on "when and if". For the moment he remains all-powerful. He is also infinitely wise, and it is therefore most unlikely that he would ever create an immeasurable universe which would have the effect of turning him from an all-powerful God into a God whose powers are limited, because of his inability to measure the universe that he himself created. Your argument is weak; you must be suffering from exhaustion. May I suggest you retire to your bedroom until the morning?'

We have come a long way in the last few hundred years, and the human race is freer from religious dogma and political interference than it has ever been. In the Christian faith, God invented time, or more specifically invented our universal time. There was no 'before that'. God inhabited his own temporal space zone into which he thankfully plopped us and our tiny universe. We are finite, but God's time rolls on for ever, around us and strangely disconnected from us. The biblical account of Genesis – 'In the beginning God created Heaven and Earth' – was discovered in the 1870s by the English archaeologist George Smith to have been a myth copied by the Jews from the Assyrians, ancient descendants of the Sumerians. Piecing together tens of thousands of shattered fragments of cuneiform tablet inscriptions which he had found buried under a mound at Kouyunjik, opposite the town of Mosul in modern-day north-western Iraq, Smith shocked the Christian world by proving that large parts of the first book of the Bible were taken from the Chaldean account of Genesis, assimilated by the Jews at the time of their captivity in the region of Chaldea some 700 years before Christ. The Assyrians, jealous neighbours of the ancient Babylonians, were similar to the Jews and, later, to the Christians, in their belief that the Creation was caused by one all-powerful being. The Babylonians, on the other hand, held that the Creation of the world was the result of a quasi-sexual act that took place before the beginning of time in some murky underworld. Tatvu (the sea) with Absu (the deep) sludged about in the formless mud and begat Mummu, the representation of chaos. Mummu needed no sexual partner to conceive Lahma (the force of growth) and Lahama (silt). Their offspring were Kisar and Assar, a brother and sister representing the lower and upper expanse (i.e. heaven and earth in a formless, embryonic stage). Their incestuous liaison produced more births: Anu or Ouranus is born, the patriarchal god of the sky, or heaven, as well as his sister Anatu, goddess of the earth. Needless to say these two unite and spawn four more god-like figures: Vul, Bilcan, Hea and Istar (see Table 1). This at least is one version of events, and there are many others.

Table 1:

A pedigree of the Babylonian gods showing the order of the Creation

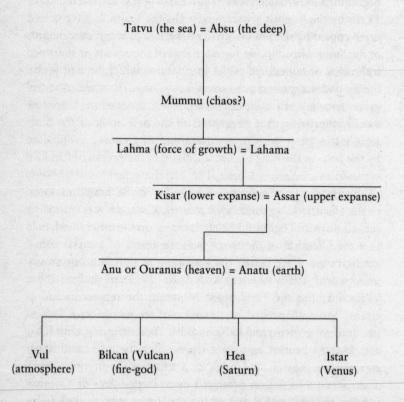

Tatvu (the sea) = Absu (the deep)

Mummu (chaos?)

Lahma (force of growth) = Lahama

Kisar (lower expanse) = Assar (upper expanse)

Anu or Ouranus (heaven) = Anatu (earth)

Vul (atmosphere)   Bilcan (Vulcan) (fire-god)   Hea (Saturn)   Istar (Venus)

In the Babylonian myth of Enuma Elish, it seems that Absu (the deep) entered into a relationship with the chaotic Mummu:

> When on high the heaven had not been named,
> Firm ground below had not been called by name,
> Naught but primordial Absu, their begetter,
> And Mummu-Tiamat, she who bore them all,
> Their waters comingling as a single body.

The Zunis, a tribe of Pueblo Indians living along the Zuni river of New Mexico, held to similar beliefs. For the Zuni folk everything started in Anon, the nethermost world, where the seed of men and creatures took form and increased 'even as eggs in warm places speedily appear. Everywhere were unfinished creatures, crawling like reptiles one over another, one spitting on another or doing other indecencies, until many among them escaped, growing wiser and more manlike.' Other myths centre on the idea of the Creation springing from a cosmic egg. One African version goes so far as to describe the impatience of one androgynous hatchling who crashes out of the egg before maturation in an attempt to control the creation process. By doing this he accidentally takes parts of the egg with him, which causes an imperfect world to be made.

All myths and religions eventually (to the scientific mind at least) fail to give a satisfactory answer to the central question: did time have a beginning or not? That is the question to which all schoolchildren are burning to have an answer – an answer that neither religion nor myth is able to provide without requiring a leap of faith, an unwilling suspension of our disbelief and a tiresome amputation of our natural organs of logic.

Socrates was, according to Xenophon, Aristophanes and Plato, one of the brightest men of his age. He was ugly to look at, maybe; statues inform us of his snub nose with grotesquely wide nostrils and a gaping mouth; he was a squat, cocky figure with a shrew called Xanthippe for a wife – this was his external lot.

To Plato, though, he was 'all glorious within' and was reported to have a sense of humour in spite of his staggering conceit. But although his enemies thought Socrates the most irritating man alive, his disciples adored him. Cicero said of him that he 'brought philosophy down from the heavens to earth'. Plato, some 40 years his junior, devoted a lifetime to recording the wisdom of his master. The importance of Socrates was not what he wrote (as far as we know that was nothing at all) or what he said – it was the way he made other people think. He used to call himself 'the midwife of other men's thoughts'. His famous trick was feigning ignorance (a technique now known as Socratic irony). By asking seemingly naive questions, he was able to squash his rivals in debate – a similar technique, one supposes, to the relentless four-year-old who keeps asking 'Why, Daddy, why? Daddy, why, why, why?' as a knee-jerk reaction to any adult postulate. That is why others found Socrates such a pain in the neck; but it also explains his importance.

Socrates was influenced by Parmenides, the Eleatic. It is recorded that they met, once, when Socrates was a young student and Parmenides an old man at the height of his fame. Zeno was one of Parmenides' star pupils and has for 2,000 years been but a footnote to the history of philosophy, for it was not until the nineteenth century that Charles Lutwidge Dodgson, better known as Lewis Carroll, the strange genius behind *Alice in Wonderland*, and later the philosopher Bertrand Russell noticed the importance of his work.

Five paradoxes are all that remain of Zeno – not written down but passed on, the first four by Aristotle in *Physics* and a fifth (possibly not by Zeno) cited by Simplicius in his commentary on Aristotle. Zeno came up with his paradoxes around about 460 BC. The first is probably the most famous:

Zeno asks us to imagine an athlete. History does not relate, but one might suspect that Zeno's own narration might have involved a description of bulging biceps, well-oiled thighs and locks of golden hair that are blown by Zephyr's soft winds on the racetrack of Olympia – for it was ever thus that the Greeks enjoyed describing their athletes. Zeno sets the problem. If the athlete is to run a race of 100 metres, how is he to pass the winning post at the far end? For to get there he will first need to pass the halfway mark at 50 metres. Once he has arrived at the 50-metre mark, he will still have an equal distance to run; yet to complete the last 50 metres he must *first* pass the 75-metre point. Wherever he is, however far from or close to the winning post, our oily friend has a remaining distance to travel which he cannot complete before he has run half the distance first. As we keep dividing the remaining distances in two, he will get closer and closer to the finish without ever reaching it – so, according to this anomaly, he can run his race, but he will never reach the end. Looked at from another perspective, our beleaguered Adonis can never even start. Inverting the same argument, we can see that for him to reach the end he must get to the 50-metre mark, but to get there he must have already passed the 25-metre point. But how is he to reach the 25-metre mark if he has not already got to the 12.5-metre mark? And so on and so on. He cannot go one millimetre until he has first gone half a millimetre. If this series of divisions is infinite (and mathematics tells us that it is), then the athlete can never start his race because there will always be one more tiny subdivision that he must reach before he can proceed to the next.

It does not take a genius to work out that there is something wrong here, but the question is: what is it that causes this to be a paradox? We all know that athletes can and do run races and that they have no problems in either starting them or finishing them – Zeno was not suggesting that they do. The fault, it seems, lies with mathematics and its relation to reality. In maths you can, in principle, divide any number higher than zero by two and get a meaningful result, but what Zeno's paradox seems to be telling us is that, in the physical world, mathematical principle does not seem to apply. If the athlete succeeds in reaching the winning post it must be that at some stage he reaches a point so close to the winning post that the remaining distance between him and it is so minute that it cannot be divided – with the happy result that he manages to reach his destination. In other words there is a physical limit to how small a space can be. Let us call this smallest unit of space a *minimum*.

Now, assuming that this *minimum* does exist (and it certainly appears to offer a rational solution to Zeno's paradox), then we can assume that as the athlete runs towards the post, the remaining distances (which are in theory constantly dividing by two) get smaller and smaller but not *ad infinitum*, for when he reaches the last *minimum* of space, no further division is possible, with the result that he is able to cross the finishing line. The implication, then, is that movement from one end of a *minimum* to the other cannot take any time, because if the *minimum* cannot be subdivided then there can be no time in which the athlete can be said to be between the two ends. Either this, or the time it takes to traverse a *minimum* must be an equally minimal, indivisible unit of time. Versed scientific minds which have bothered to follow this train of thought will have noticed that we have now arrived at the gateway of quantum mechanics and Einstein's relativity. Zeno could not have been expected to understand either of these extraordinary theories from where he was sitting in his Eleatic gymnasium all those thousands of years ago, yet the raw material was there. It was with 'thought experiments' such as these that Einstein first spawned the philosophical ideas about space and time which were to lead to his two gigantic scientific theories of relativity.

So Zeno tells us that space and time cannot regress by infinitely small degrees because, sooner or later, we will reach an indivisible *minimum*. Fine – well, almost fine. While Zeno and others are congratulating themselves on the smooth and satisfactory course from paradox to interesting conclusion, others may have spotted a problem. An even more perplexing second paradox seems to have evolved from our conclusion to the first.

Imagine a race between something spectacularly fast (let's say a photon) and something notoriously sluggish (like a slug). Light travels, we are told, at 299,792,458 metres per second – the fastest speed there is. Slugs, on the other hand, may not be the slowest movers in our universe, but they are unquestionably slower than light, taking perhaps more than three days to cover a kilometre. Now, if the slug and the photon were to race each other, even a chicken brain should be able to predict the result. But let's assume that the race is only one *minimum* long. We know that a *minimum* cannot be subdivided, so that neither the slug nor the photon can, at any moment in time, be perceived to be between the starting and the finishing posts. Assuming they both set off at exactly the same instant, they must arrive at the finishing post together. The slug cannot be slower than the photon. If they race over two kilometres – a distance of x *minima* – the result will consequently be the same. Oh dear, another paradox! Here we need quantum – in the form of Heisenberg's Uncertainty Principle – plus a decent smattering of relativity theory to help us out. But that is for later. The question that holds us here is how can Zeno's paradox help us to untangle the antinomies of time?

In the early part of the twentieth century the quantum physicist Max Planck, without reference to Zeno, established a term for the minute distance that we have been jauntily calling a *minimum*. He called it a 'Planck length'. He also invented 'Planck time' as the smallest indivisible unit of meaningful time. It had been noticed for some while that the predictable classical laws of gravity and space-time ceased to operate when observing extremely small things like atoms and particles or even less small things under extreme conditions. This observation has formed the basis of modern quantum mechanics. Extrapolating from the relative sizes of the constant of gravity, the speed of light and Planck's constant (being the relation between the wave nature and the particle energy of a proton), he was able to come up with his definition of a Planck length; described as 'the smallest measurement of length which has any meaning', it is roughly 10^{-33} centimetres, which is 10^{-20} times the size of a proton. Planck time is therefore calculated as the time it takes light (the fastest of all things) to travel the distance of a Planck length, a feat apparently attainable in 10^{-43} of a second, thus the smallest unit of meaningful time.

The most popular theory about the birth of our universe is known as 'the big bang', a babyish name coined as a term of disparagement by a Yorkshire astronomer, Sir Fred Hoyle, for a scientific theory that he once admonished for being 'about as elegant as a party girl jumping out of a cake'. The theory came about after that lawyer-turned-astronomer, Edwin Hubble, discovered in the late 1920s that our galaxies are flying apart from each other. This led to the belief that all the matter in our universe is the shrapnel from one gigantic explosion flying away from a central hot point into the great unknown. Hubble's discovery of the expanding universe marked the beginning of a whole new phase in cosmology. Before that, the universe was considered to be eternal and unchanging. This idea had settled in the brains of men as deeply as an ingrained universal truth can settle – so much so, in fact, that Einstein, working on his first theory of relativity in 1911, when he produced an algebraic sum that showed the universe to be expanding, in disbelief added an extra term to his equation to steady it – an action he later described as the 'biggest blunder' of his entire career.

Hoyle is not alone among dissenters. Professor André Linde of Stanford University, California, accepts the principle of the big-bang theory but has put forward his own version showing that our big bang might have formed just a fractional part of the act of creation. Unveiling a computerised simulation of his new theory in January 1999, Professor Linde explained:

> Each of these peaks is a huge new part of the universe, which you can consider as the beginning of another Big Bang. From the point of view of the general geometry of space, our part of the universe has been created, but other parts of this same universe are still being created. If life in our part of the universe were to disappear, then it will appear someplace else. So the universe as a whole becomes immortal.

This is advanced thinking. Most scientists are currently holding with the theory that the universe formed from a single helium explosion some 15 billion years ago. The date is approximate but has been gleaned from a careful study of background radiation in outer space and the generous sharing of information between astronomers and physicists. Scientists believe they have an understanding of the conditions that existed from 0.0001 of a second after the moment of creation. Everything that existed was squeezed into a radioactive fireball with a density 100,000 billion times that of water, and it was hot – very hot: 1,000 billion degrees above absolute zero. In these conditions particles of light become so energetic that they are able to convert themselves into particles of matter, swapping energy for mass, as Einstein demonstrated was possible in the most famous of all mathematical formulations: $E=mc^2$, where E stands for energy, m for mass and c for the speed of light (Einstein's universal constant).

Lee Smolin, a visionary theoretical astrophysicist from Los Angeles, suggests that the universe might have been evolutionary, developing and correcting itself along Darwinian lines. If this is the case then a proton now might not behave like a proton used to behave 15 billion years ago and if this is true, then much of the commonly accepted data that has been collected by scientists concerning the age and development of our universe may in fact be wrong. Back to square one!

But for those who hold with the traditional view, Lee Smolin and his ilk are an irrelevance. The question they would like answered is: what happened before that 0.0001 of a second after the big bang? Here we get into the mysterious world of 'singularities' and this, once again, is where the normal laws of physics inevitably break down. A singularity is a point described by Stephen Hawking and others as possessing infinite density (in this we presume him to mean mathematically, not physically infinite – scientists are careless with words). Such points come into being as the result of the enormous gravitational forces of black holes in space. In brief, a black hole is a collapsed star; the relation of its mass to its density has reached a critical limit so the star becomes overwhelmingly dense and hot, with its corresponding force of gravity so strong that nothing, not even light, is able to escape from it. Everything that exists – time, space, energy and matter – is wrapped around it and squeezed down to the size of a single, invisibly tiny dot. This is what is meant by a 'singularity', and it is widely believed that the big-bang explosion from which our universe was created derived from just such a singularity over 15 billion years ago.

When we consider the meaning of time, it can be explained only as a by-product of events. One thing happens, then another, so only by two events occurring can a formulation of time come into being. Einstein described things as 'events' because a thing (and by this he means absolutely anything from a particle to a policeman) exists in only one form in a particular place at a particular instant in time – at no other instant of time was it or will it be *exactly* the same again, for all things are changing constantly. In this way Einstein could maintain the clear notion that 'things' are indeed 'events'. Events, and only events, are what give us a sense of 'before' and a sense of 'after'. Similarly, duration (or time lapsed) has to be understood in terms of events – there is no other way. Time can be measured only through the medium of events and so, naturally, if there were no events, time would simply cease to exist. Equally, if there were only one event, anywhere in the whole universe and nothing else outside it, time would have no conceivable meaning.

A singularity presents just such a possibility. To ask what happened *before* the singularity is a nonsensical question. There can be no 'before', nor any 'after', nor even a duration. At the point where a singularity comes into being, time just ceases to function. To a philosopher the very idea of a uniquely single object surrounded by nothing at all dictates that this must be the case. Philosophers are supported in their 'thought experiments' by Einstein's algebra for *Special Relativity*, which demonstrates conclusively that time can be warped by the effects of gravity to such an extent that, in the extraordinary conditions of a singularity, it would simply vanish from being.

There is no point in suggesting that a singularity might be only a few billion light years from a star. The effects of a singularity are such that space and time are stopped; there can be no relative distance from a singularity to any other point outside itself as it is a closed universe. To say that our universe was created from a singularity is therefore not completely accurate, since there is no meaningful 'before' or 'after' where singularities are concerned. All we can say is that time began roughly 15 billion years ago and exactly one unit of Planck time (or a Zeno *minimum*) into its creation. In other words the universe came into being when it was already 10^{-43} seconds old. Another wretched paradox? Not really, because Zeno taught us that neither time nor space, those two inseparable dimensions of being, can be divided beyond a certain point. Zeno's fancy athlete arrives at the finishing line of his race in just the same way as the first unit of time springs into being from the timeless void of a black hole singularity.

Another simple, non-mathematical way of telling that the universe, and therefore time itself, must have had a beginning lies in a solution that Edgar Allan Poe found to the long-running debate over Olbers' paradox. Poe (1809–49), admired for his ghoulish, Gothic tales and his fine poetry, was also something of an amateur scientist and, a year before he died, wrote a prose poem entitled *Eureka* in which he explained an ancient mystery. Heinrich Olbers, a nineteenth-century German astronomer, was not the inventor of his paradox – someone else did that – but by wrongly suggesting a solution to it in 1823 he broadened the debate, and the paradox was called after him.

What Olbers' paradox suggests is that in an infinite, unchanging universe (such as mankind believed it inhabited in Olbers' time) there should be no night sky. How can a dark universe be full of an infinite number of bright stars? If the night sky is really as we believe it to be, an infinite space filled with an infinite number of bright stars, then how come it is dark when we look at it? Surely, if space is really infinite then it should twinkle with brightness in every direction, for whichever way you look your line of vision should intersect with the brightness of a star. Johannes Kepler, the brilliant seventeenth-century scientist who discovered the laws of planetary motion which led to Newton's law of gravity, had toiled over this paradox too. By the end he was quite mad, had rotten breath and thought himself to be a dog. Newton and Halley had also puzzled over Olbers' paradox (though it wasn't called that then) but the obvious answer, one that was staring them in the face, remained unseen until Poe explained the phenomenon of the dark gaps between stars which he called 'voids':

> The only mode, therefore, in which we could interpret the voids which our telescopes find in innumerable directions, would be by supposing the distance of the invisible background so immense that no ray from it has yet been able to reach us at all.

To Poe this was a gigantic discovery. Eureka! But sadly, nobody paid him much attention – after all, he was a poet and a writer of Gothic short stories, not a scientist. Why should he know anything?

Kelvin had skirted the same solution in 1904, and a follower of his, the Irish scientist Fournier D'Albe, spelt it out even more succinctly in 1907: 'If the world was created 100,000 years ago, then no light from bodies more than 100,000 light years away could possibly have reached us up to the present.' The answer to Olbers' paradox lay in the simple fact that light takes time to reach us. Looking at the darkness between the stars of a night sky should have been enough to inform our ancient ancestors that the universe had a definite beginning.

So much for stargazing science. Before that, scholars attempted to calculate the age of our universe by using history, the Bible, Aristotle and any scriptural texts they could lay their hands on – and a very unreliable method that turned out to be.

One scholarly scheme consisted of counting backwards from Jesus's time through Roman and Near Eastern eras to the ages of Abraham and Isaac, who begat Jacob the father of Judah, whose son, Perez, begat Hezron, and so on and so forth, laboriously marking each generation of this spectacularly fatuous family tree as a period of 30 years and then making the odd 'leap of faith' just when the trail goes faint and the Bible gets hazy. This method usually offered some ballpark figure that could then be rounded up, or down, to the nearest 1,000 years in order to suit some fanciful millennium-creation theory. One of God's days, as we shall later see, is reckoned as 1,000 years.

In the Septuagint, a translation of the Bible undertaken by 72 Jewish scholars (six from each of the 12 tribes of Israel), there is a calculation that puts the Creation of the universe at 5,500 years before Christ. Originally a Hebrew document, the Septuagint was later used by the Christians to lend credence to certain Christian arguments; this enraged the Jews so much that they eventually abandoned the Septuagint altogether. According to the Christian reading of this sacred text, Christ was born not on the turn of a millennium as some had thought, but right in the middle of one. If God created man on the sixth day then Jesus, the beatific symbol of mankind, must have been born in the middle of the sixth millennium, so the Creation could be placed with historical confidence 5,500 years before the birth of Christ. Others worked it out differently. The Venerable Bede, for instance, dated the Creation 3,951 years before Christ, arguing vehemently that 49 years (seven times seven years) had elapsed in human terms between God's creation of light out of the void and his completion of the process on the seventh day. Therefore (and this is typical of the bizarre logic of the medieval theologian), one needs to knock off 49 years from the 4,000 which Bede claimed others had postulated as the date of the Creation, in order to arrive at the true date of the Creation: 3951 BC. Of the many theories that abounded, perhaps the most successful, or at least the one that gained the widest approval, was a calculation by the Irish divine, Archbishop James Ussher. Ussher was head of the Anglican Church in Ireland, and it was he who published in 1656, in his chronology *Annales veteris testamenti a prima mundi origine deducti* (*The years of the Old Testament, calculated from the First Origin of the World*), the view that the Creation of the world took place in 4004 BC, precisely at midday on 23 October! The archbishop pointed out that Herod the Great was known to have died in 4 BC and that Jesus, or so the story goes, was born in the reign of this Herod. Jesus must therefore have been born in or before 4 BC. Ussher's views were hugely popular throughout most of the Christian world, where 4004 BC became widely accepted as the official date of the Creation. Which was not to say that Ussher, or any of the others for that matter, believed that time itself started then.

Time was there all along. What God did was to place the Creation of the universe into his pre-existing mould, which was made from eternal time. St Augustine, who was as obsessed by time as he was by his mother Monica, believed that God actually fashioned our universe out of time, using it as a raw material. Leibniz, the seventeenth-century German philosopher, in another piece of utterly muddled thinking, saw God's Creation of the universe as a logical reason for the world to be either non-existent or eternal:

> For since God does nothing without reason and no reason can be given why He did not create the world sooner, it will follow either that He created nothing at all, or that he created the world before any assignable time, that is, that the world is eternal.

It beggars belief that a man of undoubted genius, a brilliant mathematician and such an inspired all-rounder as Leibniz could have wasted his brain, ink and his paper on such a shallow consideration. As he wrote these words he probably thought he was being irrefutably clever.

Must we all suffer these posthumous indignities? Will future generations spot the flaws in our own reasoning, guffaw at our foolishness and chide us for our arrogance? Probably!

Chapter II

MOMENTA
Seconds

The debt we owe to the Sumerians is incalculable. They marked the beginning of human civilisation as we know it, introducing a much-needed element of calm into the frantic maelstrom of ancient life. They built cities, great cities: Kish, Erek, Ur and Sippar were among them. Each was walled, representing a state, with land and villages around it. In the centre, a temple provided a site for meeting and worship. This all sounds balefully familiar to us; history has unveiled thousands of cities that have risen and fallen, have thrived with hope and initiative at some era or other and subsequently squabbled themselves to ruin. To the Sumerians, though, this was all new, new, new! There is no record of any civilisation before it. It was, as far as we can tell, the beginning of modern times.

Who the Sumerians exactly were is uncertain; their origins were obscure. The area from which they came has changed its name through history: from Sumer to Babylonia to Mesopotamia and is now the part of southern Iraq that stretches from Baghdad to the Persian Gulf. Sumer was first populated by a non-Semitic race who did not speak the Sumerian language. These people, called proto-Euphrateans or Ubaidians, were not complete barbarians. They were responsible, after all, for rudimentary agricultural practices, draining the marshes and developing trade. Pots, metalwork and weaving from this period indicate an average talent, an eye for modest decoration and a will to self-improvement.

The people we now refer to as Sumerians were immigrants to this region who brought their own language with them, a language which took over from whatever the Ubaidians were talking. These new people instigated the remarkable blossoming of civilisation in the area. If the non-discovery of earlier examples can be counted as evidence, then we are in a position to assert that the Sumerians invented not just walled cities and city-states but vehicles with wheels too; they also invented potters' wheels, board games, ball games, the first system of writing (known as cuneiform), the first codes of law, written mathematics, geometry saws and treadmills, swings and hammocks, calendars based on lunar phases and, in a roundabout sort of way, the second.

The word itself is not Sumerian. It comes from the Latin *secunda*, meaning 'second', as in 'coming after the first', and being a diminution of *secunda minuta* or 'second minute'. The 'first minute' or *prima minuta* (that which we now call a minute) represents the first operation of sexagesimal division (i.e. dividing the hour into minutes); the second or *secunda minuta* (that which we now call a second) is the next operation of sexagesimal division (i.e. dividing the minute into seconds). We do not know if the Sumerians had a word for the chronological second. If they had, one is tempted to speculate that it might have been *gesh*, a word they also used to mean 'man' and 'erect phallus'. No hasty conclusions! The word *gesh* was also used by these people to signify the number one, and as a consequence was employed to describe one geometric degree ($1/360$ part of a circle) in their unique system of counting and geometry based on the number 60.

Why was the number 60 so important to the Sumerians that, 7,000 years on, our modern minutes and seconds are still based on it? No other civilisations were so concerned to count in base 60. In many ways it is inconveniently high. The Romans, who worked principally in the decimal system, as we do, also had weights and measurements based on the duodecimal system of base 12. They had, for instance, the *as*, a unit of arithmetic, weight and money which divided into 12 ounces. Even now duodecimal systems are still in place. In Britain and America we buy eggs and oysters by the dozen, we still talk about a gross (144, i.e. 12 times 12). The French changed to the metric system after their revolution of 1789, and the rest of the world is still catching up slowly. But before the French dived so enthusiastically into their *système métrique*, their common coin, the *sol*, was divisible into 12 *deniers*, and in the imperial system, still used by old-fashioned types in Britain and America today, the foot is divided into 12 inches. In the old French version the inch could be subdivided into 12 equal lines and the line into 12 equal points.

The duodecimal system is responsible, as we shall later see, for there being 24 hours in a day, but why the Sumerians favoured the sexagesimal system over this, the decimal or any other system has been hotly debated. Perhaps it was simply the easiest to use, being the lowest of all numbers with the greatest number of divisors. In other words it is the smallest denomination of coin which you could take to your local market and say, 'For this 60-*gesh* coin I am able to buy from you either one bull, two cows, three sheep, four pigs, five dogs, six undergarments, 10 hammocks, 12 treadwheels, 15 of those funny calendars, 20 wheels, 30 wooden lavatory seats [yes, the Sumerians had those too!] or 60 brown eggs. I have 12 purchase options. Luckily, I have just found another 11 coins in my pocket. I will take the lot – no change needed. Byeeee!'

It was with this market-convenience argument, or one very like it, that the fourth-century Greek scholar Theon was able to explain the Sumerian 60. If this were actually above the intellect of the Sumerian businessman (we have no reason to suppose it was) he would at least have noticed that 60 was the lowest number into which the first six integers divided, and this in itself would have made the Sumerians consider the number as something rather special.

Another view, which was first proposed by the mad Venetian Formaleoni at the time of the French Revolution, suggests that the Sumerians were particularly interested in astronomy. (NB: It has been common among historians of recent ages to ascribe to the peoples of more ancient times an all-consuming passion and sometimes a mystic understanding of the stars. This is not always to be believed, as we shall later see in our review of the builders of Stonehenge. There is evidence that the Sumerians learned to count time by the phases of the moon, but this is not to be confused with either an obsession or even a religion based on the spheres.) Formaleoni's view was that the Sumerians, having calculated 365 days to the year, for some reason (and this is not explained) conveniently rounded it down to 360. If the year consisted of 360 days, the Sumerians (again according to Formaleoni) would have taken it on themselves to divide circles equally into segments of 360 degrees. This would permit each drawn segment to represent one day and the whole circle to represent the year. One thing the Sumerians did discover was that the chord of a sextant (one-sixth of a circle) measured equal to the radius. Whether or not they held the circle to be a magical shape and 60 (by virtue of its relation to the circle) to be a magical number is not recorded. Perhaps the link between six, 60 and 360 and circular geometry encouraged the Sumerians into implementing 60 as a useful unit for counting pigs at the market; but the argument is not altogether a logical one. In spite of its weakness this view succeeded in gaining the support of a famous historian of mathematics, Moritz Cantor (1829–1920) – not to be confused with the German mathematician of Danish extraction, Georg Cantor (1845–1918), whose particular fame rests on a daring theory of infinite numbers which seems, on close inspection, to be even more illogical than his namesake's reflections on the Sumerian counting system.

Another view, similar to that of Theon, suggests that the Sumerians changed their system from base 10 to 60 in order to harmonise measures of weight corresponding to measures of price or value. That one third or two thirds would not divide into 10 was evidently inconvenient; even so, the likelihood would appear that the Sumerians did not change their system of mathematics for convenience but that their weights and measures were designed to suit their system of mathematics.

The reason for the Sumerian 60 which we would most wish to be true is the exciting one, the mystic one, the one in which ancient secrets are revealed and the meaning of the universe is unlocked in a timeless riddle to which 60 is the answer and the grail. The Sumerians believed that the number 60 had magical powers – did they?

There are certainly connections between numbers and gods which crop up in those later Mesopotamian civilisations which orginated from the Sumerians. The most magical number of all, 3,600, was called *sas* or *saros*, its cuneiform symbol being a contorted circle; the word was also used to mean 'everything', 'whole' and 'cosmos'. Assar and Kisar, the incestuous Babylonian gods, previously described as representing 'upper and lower expanse', who came together to give birth to Anu and Anatu, take their name from the number 3,600. Anu himself, the god of heaven, is represented by the number 60. In cuneiform, the symbol for 60 is the same as the symbol for Anu, and our old friend *gesh*, the word for man, erect phallus, etc., was, as previously stated, also used to denote 'one' or 'one degree' (i.e. $1/360$ part of a circle). Between Anu and *gesh* we have a whole host of numbers relating to celestial and earthly beings. The sun-god and 'king' is 20; the moon-god is 30 (possibly by virtue of the number of days in a month). Higher up the gods' ranking system, Abyss is allotted 40 and Anatu, the incestuous earth-god, is 50.

Once again the most likely inference is that the Sumerians did not create their system of numbers out of their mystical beliefs, but the other way around. Anu, the great god of the heavens, was given the number 60 because 60 was the most important number in the Sumerian counting system and Anu the most important god.

By far the most likely reason for Sumerian sexagesimalism has been put forward recently by a French mathematical historian, Georges Ifrah. Most number systems seem to have originated from finger-counting. Interestingly, the Sumerian words for 'six', 'seven' and 'nine' mean, respectively, 'five plus one', 'five plus two' and 'five plus four', which implies that their ancestors (were these the averagely talented proto-Euphrateans?) were using a number system based on five; the Sumerian word for 'ten', incidentally, *u*, also means 'fingers'. It might well be that the Sumerians drew from the itinerant peoples around them and that a system combining bases five and 12 (five times 12) resulted in the uniquely Sumerian base 60. Georges Ifrah constructed a plausible system of finger-counting which may well be the origin of Sumerian mathematics and thence the origin of our minutes and seconds.

Look at your right hand with the palm facing. You will (unless there is something peculiar about your hand) notice that each of the four fingers is distinctly divided into three articulations or phalanges. Might the Sumerians have counted these phalanges? Since these people were built the same as we are, let us imagine, as we inspect our beautifully clean modern digits, that we have before us the primitive paw of a mathematically inclined Sumerian. If you count the phalanges, they should come to 12. In much the same way as an abacus works, the fingers of the right hand can be used to count single numbers from one to 12 while the fingers of the left mark off the twelves. The little finger of the left hand, let's say, represents 12, the next 24, the long middle finger 36, the left index 48 and the thumb 60. There you have it. The fingers, once you get used to working them, form a perfectly adequate abacus in base 12, of which the highest number is 60. By this means, any number lower than 60 can be easily shown. Thirty-four, for instance, is represented as 24 plus 10 (24 by raising the first two fingers of the left hand, and 10 by touching the top phalange of the right-hand index finger with the otherwise redundant right-hand thumb). Try it! Ifrah admits that the hypothesis is, as yet, unproven, but he points out that 'phalanx-counting of this type does exist and is still in use today in Egypt, Syria, Iraq, Iran, Afghanistan, Pakistan and some parts of India. Sumerians could therefore easily have used it at the dawn of their civilisation'.

That the Sumerians divided their circles into 360 degrees is uncontested. Their hours, however, known as *danna*, were the equivalent to two of our modern hours, which gave 12 Sumerian hours to the day. Minutes and seconds were all part of the plan, on paper at least, but there is no evidence to show how they might have used them.

In science, we know that seconds of time (as opposed to geometric seconds, or degrees as we call them) were used as far back as the second century BC. The Greek astronomer and mathematician Hipparchus (d. *c* 127 BC), famous for his discovery of the precession of the equinoxes (even though it was probably not he but the Babylonian Kiddanu who discovered them), managed to calculate the length of the year to within six and a half minutes.

To the average Sumerian, seconds, we can safely assume, did not impinge on the daily run of affairs. Even in our own times seconds are hardly used by the ordinary citizen. To those not involved with science the second has never been of any great use, though people have, for some time, spoken about them in a vague, non-literal sense simply to mean an extremely short time (for instance in the phrases 'just a second', 'in a split-second he was out of the door' or 'I'll be with you in just a tick'). The word, used in this loose sense, was popular throughout the nineteenth century. Earlier, Congreve, in his play *Love for Love* (1695), implores Sir Sampson Legend to be on time: 'At ten O'Clock, punctually at ten, Sir Sampson. To a Minute to a Second; thou shalt set thy Watch and the Bridegroom shall observe its motions.' The earliest known written English use of seconds as units of time dates from 1588, in a barely comprehensible translation from a Latin treatise by A. King: 'Ye cowrse of ye sone, quhilk sence has bene obserueit to be accompleseit in 363 dayes 5 houris 10 min: and 16 secondis.' Or, to retranslate: 'Bright minds (*sic*) have noticed that the duration of the sun's orbit (*sic*) is 363 days, 5 hours, 10 minutes and 16 seconds (*sic, sic, sic*).'

Timekeeping in the Middle Ages and earlier Dark Ages was, throughout Europe, the jealously guarded secret prerogative of the Church. It was the Church, after all, which annually calculated and proclaimed the date for Easter. The standard definition for this date (i.e. 'the first Sunday after the full moon (paschal moon) that occurs upon, or next after, the vernal equinox') was clearly out of the range of normal laymen to interpret. And so the measuring of time became the honoured occupation of the Church. Chief among the monks and holy scholars who studied it was the Venerable Bede. Two of his volumes, *De temporibus* (*On Times*) of 703 and the longer version, *De temporum ratione* (*On the Reckoning of Time*) of 725, were mainly concerned with the calculation of Easter and formed part of a literary genre then called *computus*. Bede's books found their way into all of the great libraries of Europe in the 600 years following his death.

We do not know who Bede's parents were, nor exactly where they came from, but the holy monk himself was a Jarrow man. Shipbuilding had not begun at Jarrow in the first millennium; at that time it was a small farming and monastic community in the Northumbrian coastal district of north-east England with an estimated population of 300 adults. Bede's *computus* texts, when they are not concerned with the question of Easter, deal explicitly with time duration, from the largest units (*era*) to the very smallest subdivisions of seconds (*ostenta*). The disarray of medieval thought on this matter is plain from Bede's at times exasperated work. He records, for instance, a particular set of time divisions in which the day is divided into 12 hours and each hour into 30 *partes*; each *parte* can be broken down into 12 *puncti* and each *punctus* was the sum of 40 *momenta*; and each *momentum* is then deemed to constitute 40 *ostenta*. A quick tap on the calculator and we can show the modern equivalent. An hour (two hours), a *parte* (four minutes), a *punctus* (20 seconds), a *momentum* (half a second) and an *ostentum* (about $1/100$ of a second). Also in *De temporum ratione*, in the chapter entitled *De minutissimis temporum spatiis* (*Of the smallest units of time*), Bede rejects the notion that these minuscule units amount to any more than theoretical posturing. In particular, he ridicules one suggestion that the hour should be, or is, divided up into 22,560 atoms, $^{16}/100$ of a second each. Bede records neither who it was, nor how the calculation of 22,560 atoms of time to the hour came about, but he is nevertheless highly critical of the whole idea.

The notion that all matter was made up of discrete atoms orginates, as far as we know, from the ancient Greek philosophers Leucippus and Democritus. Later writers, such as Epicurus and the poet Lucretius, supported the theory and it may well be that Bede had in front of him an ancient Greek text, or at least a transcription, which proposed that time itself was not a continuous fluvium, but a series of atomic instances (just like Zeno's *minima*). But why 22,560 of them to the hour? We shall probably never know.

And so it would appear that seconds, minutes and hours were not standardised at the time when Bede was writing in the eighth century AD, and yet the purpose of his treatise was to point out – long before the invention of the mechanical clock – that something must be done about it. Bede backed the division of the day as we now have it (24 hours of 60 minutes each and 60 seconds to the minute). He was not alone in this. Others saw that such divisions were not only practical but were rooted in the ancient geometry of Euclid and Pythagoras. It was the job of theologians like Bede to show how the teachings of the ancients, in particular Aristotle and Euclid, were in harmony with modern Christian thinking. Yet it stands as a monument to the organic nature of human development that the time divisions of the day, which are currently used all over the world, were never officially decreed. There was no royal edict or papal bull insisting that the hour should consist of 60 minutes and the day of 24 hours. It all happened as though by stealth. Minutes and seconds must originally have been devised as theory, calculated as a parallel to the ancient geometric divisions of the circle into 360 degrees. It was only a matter of time until the scholars of the medieval Church in the late eighth century, encouraged by the writings of Bede, had arrived at a general consensus that divisions of the hour should likewise be based on the radix 60. By the time the first mechanical clocks were made, some 400 years later, what was once a general consensus had now evolved into established fact. The Christian Church was, after all, entirely responsible for these matters. To the average citizen, minutes and seconds would have seemed whimsical and oblique, yet the symbolic convenience of the circular clock face corresponding to the ancient geometry of the circle would not have escaped the notice of the medieval theologians to whom these matters were of the utmost importance.

If seconds were established, at least by consensus, some time after the death of Bede, they remained a backwater of theological theory for many centuries to come. Since they were impossible to measure with any degree of accuracy, seconds were of no practical value to the people of the Middle Ages.

Sundials and shadow clocks were never accurate enough to identify time in units as small as seconds. It was not until the late sixteenth century when the Italian mathematician, astronomer and physicist, Galileo, started experimenting with gravity, that the accurate measurement of small units of pulse was at last made possible. Galileo pointed out that a heavy object dropped from a great height would take the same amount of time to hit the ground as a light object dropped from the same height, so long as the wind resistance of the two objects were the same. He also showed that a cannon ball dropped from a height of one metre would hit the ground at exactly the same time as a cannon ball fired horizontally from the barrel of a cannon one metre high. These findings seemed, at the time, to defy common sense, and yet they were true. In 1583, Galileo recorded the movement of a pendulum by comparing a swinging lamp at the cathedral in Pisa with his own heartbeat. Soon he was able to make an important discovery – that the period of a pendulum (the time it takes for it to swing from the top of its arc and back again) is not dependent on the weight of the bob but on the length of the shaft to which it is attached. The shorter the shaft, the shorter the period, irrespective of the mass of the bob. Once the theologians of the Middle Ages had, by stealth and persistence, established the theoretical time value of the second (being the sixtieth of a minute, being the sixtieth of an hour, being one twenty-fourth of a day), it was only a matter of time before men like Galileo and Christian Huygens from Holland were able to define the second with scientific accuracy as a period determined by a pendulum shaft of a set length. Thus, throughout the seventeenth century, clockmakers were able to work on the basis of a second being the period of a pendulum with a shaft length of 39½ inches.

John Harrison (1693–1776), a Yorkshire clockmaker whose name has been immortalised by the popular story of his relentless determination to work out a method for calculating longitude, was fully aware of the need for accurate second-counting. It took Harrison nearly 40 years between the construction of his first chronometer of 1735 and receiving the final payment of a £20,000 British government prize, offered for a solution to the longitude puzzle. Harrison's ingenious No. 4 marine chronometer was accurate enough to stay within five seconds of correct time on a sea voyage from Britain to Jamaica in 1762. This level of accuracy enabled the ship to calculate its position to within 1.25 degrees of longitude – a marine innovation that saved thousands of lives at sea over the coming years. In his pursuit of such deadly accuracy Harrison attempted many methods before he settled on his No. 4 chronometer. In one experiment he devised multi-rodded pendula made of different metals to compensate for heat expansion. When a conventional pendulum expands due to heat rising by only a few degrees, the clock will lose time and clearly prove inefficient on a ship in tropical climes. Harrison made two matching clocks and put one in his modest sitting room with a blazing fire and the other in the hallway with the door open on a winter's day. By running back and forth between the two he was able to monitor the effectiveness of his new system to within one degree or half of a second. Harrison's most efficient timepiece kept, it is said, to within 0.05 of a second each day.

Nowadays ordinary people go about their daily business quite oblivious to the passing of seconds; and yet we have, by preference, second hands ticking away on our watches, in motions of stark, quartz-modern precision, full of slick and status, signifying nothing. We seem to think it quite normal when the radio beeps its sine-wave seconds at us every morning or when the telephone speaking-clock articulates the time of day in seconds and Linford Christie wins a medal by a margin of a hundredth of a millisecond. Yet there seems to be a strange human hypocrisy at work here. On the one hand, in Western social terms strict punctuality is considered a failing, yet at the same time digital accuracy in our watches and clocks is hailed as an heroic quality. Indeed, to the Swiss, accuracy and precision are considered national virtues: the Swatch, the Rolex, the Omega and a thousand other brands of watches and clocks have trumpeted Switzerland as the chosen capital country of the microsecond. Switzerland, too, is the mighty home of William Tell, a fifteenth-century crossbow marksman who, by shooting an apple from his son's head, freed the Swiss people from Austrian tyranny and set the bench-mark standard of accuracy for generations of cuckoo-clockmakers to come. Tell is still a hero over there, but the Swiss love of accuracy does not give the whole picture. In Western society, even in Switzerland, it is considered rude to arrive at a dinner party on time; 10 minutes to half an hour is respectfully late. A fancy invitation to a Paris ball (the sort of thing that clutters up one's letter box every morning in high season) might have 8 p.m. embossed on it, but anyone who turns up before 10 p.m. will find he cannot get in. In large parts of the world nobody does anything on time, and yet we all go around with watches on our wrists proclaiming the exact seconds of the day. Why do we wish to carry about, strapped to our person, this absurdly accurate chronometrical information? Do we even know what a second is?

If we call a second one-sixtieth part of a minute we have not really defined it; all we have done is to ascribe to it a relative value to the minute which, in the seemingly nebulous world of time-measurement, is even less helpful when we consider that our understanding of what a minute really means is just as shaky and relative. As rank amateurs we might attempt the following definition: a year, we know, is measured against something non-relative (i.e. something outside of our own measuring system), for it is the time it takes for the earth to orbit the sun. If this is the case, then the year is a useful standard for time-measurement, far more useful than its 'relatives', the minute and the second. Most years (excepting leap years) contain 365 days, the day is 24 hours, hours each contain 60 minutes, and minutes are the result of 60 passing seconds. Right! That's 60 times 60 times 24 times 365 (be careful not to stick your tongue out as you work out this sum – a boy at school did that and was much resented) . . . Bingo! 31,536,000! That means we have at last defined the 'second' as something meaningful: $1/31,536,000$ of a normal year.

'No, no, no, no,' goes up the cry of the learned scholar, Adviser-in-Chief to the International Committee on Weights and Measures, 'you are out by 20,925.9747 seconds. If you think that is how long a second is, your year will be five hours, 55 minutes and 30.232 of my seconds out of whack.'

There are a huge number of considerations that need to be brought into play when calculating the number of seconds in a year, and consequently the second has received a number of 'official' definitions, most of them in the twentieth century. For many years the accepted definition was $1/86,400$ of a mean solar day (i.e. the average period of rotation of the earth on its axis relative to the sun). This was later modified. One can assume that the vituperative scholar above was speaking at some time between 1960 (when the General Conference on Weights and Measures ratified the definition as $1/31556925.9747$ of the length of the tropical year) and 1967, when a yet clearer definition was agreed.

There was a breakthrough of sorts in 1948 when the first atomic clock was developed by the National Bureau of Standards in America. It was noticed that nitrogen atoms, when combined with ammonia molecules, oscillated back and forth at a steady rate of 23,870 oscillations 'per second'. Already we see a hitch. If ammonia-soaked nitrogen atoms are the most reliable micro-oscillators, then what does the 'per second' mean in this context? Presumably, the second as measured by the previous most accurate machine, the quartz clock (invented in 1929), but even this must have some relation to the mean solar day or the tropical year. The ammonia clock was superseded in the late 1960s by the cesium clock, still in use today, and it was only in 1967 that an agreement was reached to define seconds, not by years or days but by atomic oscillations of cesium. Thus, the new definition became: '9,192,631,770 oscillations of the electromagnetic radiation corresponding to a particular quantum change in the superfine energy level of the ground-state of the cesium-133 atom.'

The cesium clock is accurate to one part in 10,000 billion or one second in 316,000 years. But the story is not over yet, and accuracy fanatics are certainly not downing their tools on the strength of the cesium clock's measly performance. Laser clocks, which use hydrogen atoms instead of cesium, one of which is said to inhabit the US Naval Research Laboratory in Washington DC, are believed to be accurate to within one second in 1.7 million years. Using Einstein's principles, it should be possible in the future to construct a clock whose oscillations are regular to the tune of one second in 300 million years.

What gobbledegook would this have seemed to the bright Sumerian merchant busily inventing his wheel and his lavatory seat so many millennia ago. Even to us with our clear modern minds, numbers like 9,192,631,770 have only the barest semblance of meaning, for science has propelled us into a world beyond our capacity to comprehend. We have been accelerated forward at such a pace that we now bandy these puzzling figures around as though it were all quite normal, as though we really understood what they meant.

'Ah, yes.' We nod our heads. 'Nine billion, one hundred and ninety-two million, six hundred and thirty-one thousand, seven hundred and seventy oscillations of electromagnetic radiation, eh? Funny, I thought it would have been seven point two, five, six, one, eight, three, two, one, six per cent of an oscillation more than that. How interesting!'

'And there's nowt so queer as folk,' as our old Lancashire saying goes.

CHAPTER III

MINUTAE
Minutes

H istory, which Henry Ford described as 'more or less bunk', is not very good, it must be said, at shedding light on the things that people did not know. It is all very well a history that informs us that Copernicus (1473–1543), the Polish astronomer, knew that the sun was at the centre of our solar system. It is an interesting point of obvious historical value; but an equally interesting point of equal historical value would be to know when it was that the general public became aware of Copernicus's point. When, in other words, could it have been said that Copernicus's conceptual revolution of the solar system formed part of the general knowledge of the people of Europe, as it does today?

In ancient Greek times, people who knew (the philosophers) set themselves as a caste apart from those who did not (hoi polloi). It is all very well saying that the ancient Greeks knew that the square of the hypotenuse on a right-angled triangle is equal to the square of the other two sides. Pythagoras knew that for sure, but how many other Greeks could have told you that in 500 BC? That the Old Testament warns 'Be not ignorant of any thing in a great matter or small,' has not had any effect. Sixty-four per cent of children between the ages of eight and 16 tested in a British schools' survey in 1998 were unable to name the Prime Minister of their own country. In America a staggering 83 per cent of the adult population is unable to name three European countries without getting at least one wrong. Over the centuries, education standards have undoubtedly improved for the poorer classes, but high levels of ignorance still prevail.

That most people had no idea of the date until the beginning of the eighteenth century might seem surprising. They kept a count of the days of the week, they knew when it was Sunday and what time to go to church because the church bell summoned them, as did the cry of 'Marché! Marché!' on a market day and the horn blast to call the labourers in from the field to eat. In general, people rose from their beds at dawn and went to sleep a little after dusk. With watches and alarm clocks, modern man is far more flexible and has lost his sense of time according to the sun, for he keeps it according to the clock. What a comfortable explanation that was. How well it slipped from the pen, but none of it is true, not a word of it.

The ancient Babylonians had sundials, not just for the aristocrats to use but for the whole busy, thriving city of Babylon, which was organised according to the clock. For Seneca, writing at the time of Jesus, the clock was a folly, which had encouraged the people to sleep in the daytime and to show off at night. Juvenal mentions a man who asks the gods to punish a young merchant for bringing his sundial into the city. Sundials, he complains, have encouraged the Romans to eat, not when they are hungry but at ridiculous set hours during the day. Only two centuries later, Artemidorus Daldianus, an Ephesian soothsayer (whose body was supposed to have erupted in boils if ever he told a lie), wrote, or is supposed to have written, *Oeirocritica* ('*Interpretation of Dreams*'). In it is a revealing passage about clocks:

A clock signals work and duty, action and the start of business. For men keep a watch of the time in all that they do. And so if a clock should fall apart, or have cause to be broken, it means bad luck and death.

By the second century AD, then, the clock was an institution that had been used in Rome for at least 400 years. In Book VII of his *Natural History*, Pliny tells us that sunset and sunrise were the only recognised times of the day in 451 BC. A little later, midday became an official point of reckoning, and the last hour of the day was yelled out by an official each night in Rome. Lucius Papirus Cursor was the first to bring a sundial to Rome in about 290 BC. Whether it was Greek or Babylonian in origin, history has failed to record, but we can be sure that it divided the day into 12 hours and that the Romans borrowed the name *hora*, which had been in use at the time of Alexander the Great 100 years earlier. No doubt Lucius Papirus Cursor's sundial proved a popular talking point among his luncheon guests. If he bought it abroad the chances are that he did not set it up properly in his garden. If this had been the case, one can only guess how the sundial was used to laugh at Greeks and Babylonians for their quaint, arcane practices. If Lucius's sundial had worked properly, would the fashion for sundials not have caught on a little sooner?

An official, Valerius Massala, was able to bring one to Rome from Catania in 263 BC, but according to Varro it took 99 years for anyone to notice that, due to the difference in latitude between Catania and Rome, the dial failed to show the right time in its new home. Valerius lived too early to heed Ogden Nash's warning:

> I am a sundial.
> I make a botch
> Of a thing that can be done
> So much better by a watch.

Pliny states that it was not until the first properly installed sundial of 164 BC and the setting up of a public water-clock five years later that the Romans started to divide the hours of daylight.

Most of the world still took its cue from the sun. There was daybreak, midday and sunset, that was all. Until very recently the people of Java divided their day into 10 natural but vague and unequal subdivisions. In order to discuss a past event they would point to the place in the sky where the sun had been when the event occurred. In Swahili, when a shadow from their body measures nine feet they say, 'It is nine o'clock', which is peculiar, not to say inaccurate of them. When the shadow is only two feet long they say, 'It is two o'clock'. In Iceland each house would choose, according to its surroundings, certain markers within the range of vision: a hill, maybe, or a neighbour's chimney stack. If there were none of these things in sight, they would make great piles of stones. These were known as *dagsmork* ('daymarks'), and the time was ascertained by the position of the sun relative to the *dagsmorks*. It meant that the Icelanders as a whole kept irregular hours, but at least each isolated farm could keep note of a time that was meaningful to itself.

Measuring time by the shadows is an ancient form of reckoning, probably the oldest there is, but it was not until the invention of the sundials with gnomons that any real understanding of equal hours came into the picture.

One of the problems with sundials is that they are not very good at marking off the minutes, and it is for this reason, one suspects, that minutes, while of great importance to astronomers, were ignored by the public at large until the majority of them carried watches in the nineteenth century. The invention of the watch dates from long before the nineteeth century, of course. Nuremberg 1477 is the first date we are given in some histories, although it is known that Robert ii, King of Scotland, had a watch in about 1310, though we cannot be sure what this watch actually was. It certainly had little if anything to do with the Victorian pocket watch or the modern wrist watch. Kings, emperors and astronomers were the only people who owned watches in the early days. A German astronomer, Friedrich Purbach, is said to have been the first to use a watch for astronomical observation in 1500. It has also been asserted that the Emperor Charles (son of Philip the Handsome by Joanna the Mad and conqueror of Suleiman the Magnificent) had something called a 'watch' in the 1530s, though it was described by others as a small table clock. We do know, however, that watches were imported to England from Germany in the late sixteenth century, and a watch that belonged to Queen Elizabeth I (1533–1603) is preserved in the library of the Royal Institute in London.

A dispute has erupted over the question of who invented the balance-spring pocket watch. The Dutch insist that it was their man, Christiaan Huygens (1629–93), and the English are determined to prove it was theirs, Robert Hooke (1635–1703). They were both remarkable men. Huygens, regarded by some as second only to Newton as a scientist of his age, made important discoveries in physics. He was the first to notice, for instance, the rings as well as the fourth moon of Saturn by using a refracting telescope which he had built with his brother. He also discovered polarisation, the laws of collision affecting elastic bodies, and propounded many insightful theories on optics, the behaviour of light, and on wave motion. Robert Hooke, who was also an excellent architect, was every bit as ingenious a scientist as Huygens was. To his track record can be added extraordinary designs for flying machines and air-pumps (the latter interest he shared with Huygens). He was known to be a bad-tempered individual who, like Newton (with whom he engaged in several acrimonious priority disputes), was never at his ease in other people's company. Even so, he had a brilliant, visionary mind and, when the mood took, 'a fair sense of humour'. He anticipated the invention of the steam engine and came up with a law that today still bears his name, Hooke's Law (which, like Huygens' work, concerns elastic bodies). With his own reflecting telescope (the first to be constructed), Hooke discovered a fifth star in Orion and noticed, for the first time, the rotation of Jupiter. His epitaph would not be complete without a mention of his ground-breaking work on the microscope, the quadrant and the marine barometer.

The Hooke camp has better evidence than the Huygens camp for settling the dispute over the invention of the spring watch. That amiable priest, Dr William Derham (1657–1735), whose hobbies included getting into the minds of wasps and the study of bird migration, wrote an early book on the history of watches,

The Artificial Clockmaker, described in its subheading as 'a Treatise of Watch and Clockwork, showing to the meanest the art of calculating numbers to all sorts of movements'. In this book Derham accords to Hooke the honour of inventing the spring watch as well as having invented a pendulum watch at the tender age of 23, in 1658. The same date appears on an inscription that can be seen inside the closing lid of a double-balance watch presented to Charles II in that year. It reads: 'Rob. Hooke, inven. 1658; T. Tompion, fecit, 1675.'

The fair-minded modern view is that Hooke got there first, but there is no evidence to show that his spring was in the form of a spiral, and that absolutely crucial element has been ascribed to Huygens.

Long before that, Alfred the Great (he who charred the cakes) was said to have invented the wax candle-clock in or around about AD 886. Whether this amounts to much more than noticing how candles burn slowly and evenly is not certain, though it was reported, many years after the event, that special uniform candles were used which burned down at the rate of three inches (7.6 centimetres) per hour, and that six wax candles were used to mark off a period of 24 hours. Alfred was also supposed to have invented lanterns to defend them from the wind.

How, though, did the ancients divide the day into small units of time? Ptolemy (AD 90–168) was an Egyptian astrologer and cartographer who was considered to have the greatest brain of his field. So much so that his long compendium to astrology, known as the *Almagest*, took its name from the Greek *megiste* ('the greatest'). There is no doubt that Ptolemy was an extremely clever man. It is only unfortunate that his incorrect view that the earth (and not the sun) lay at the centre of the universe managed to hold sway right through until the beginning of the sixteenth century, when Copernicus gingerly overturned it. For Ptolemy, the day was divided into four quarters of six hours or 360 *chronoi*. This gave 90 *chronoi* to each quarter-day. A *chronos* was therefore the equivalent of four minutes, a practical unit for everyday use which could easily be measured on a sundial shadow-clock or a clepsydra water-clock.

Adam Bede wrote much about Ptolemy, attempting to rational-ise his work to the ways of Christian thinking. Ptolemaic *chronoi* he called *partes*, and further described Ptolemy's even smaller subdivisions, calling them *momenta*. Ptolemy divided the hour into 40 *momenta*, making each *momentum* 90 seconds. *Momenta* should therefore be seen as the closest that the ancients got to the modern minute. Bede also noted other divisions, the *minuta* (six minutes) and the *puncti* (15 minutes). Later in the Middle Ages, the hour was divided into 60 *ostenta*. But we must remember that all these divisions were mathematical: for anyone to know what *ostenta*, *puncti* or *minuta* actually were would have required an accurate clock, and clocks, as we know, did not exist for the people in Bede's time.

Nor, of course, were the people of ancient Egypt and Babylonia able to measure minutes and seconds with accuracy, and yet min-utes did exist – in theory. In Babylonia, for instance, astronomers used one *ges*, which, like the Sumerian word *gesh*, was also used to mean man and erect phallus, and one *ges* could be divided into 60 *gar*. In modern terms one *ges* is the equivalent of four minutes. There were therefore four seconds to a *gar*. Minutes in India, or the nearest equivalent to them, were called *palas*, which lasted two and a half minutes each and could be subdivided into 60 *vipala* of two and a half seconds each. The 24-hour day comprised 30 *muhala*. These Indian hours were 48 minutes long. The Chinese had an unusual set of small time units which do not, on the face of it, look at all practical. One *tshei* is equal to 14.4 minutes, and the next unit down (the *ke*) is 100 times smaller (i.e. 8.64 seconds). Closer to our own system is the Hebrew *chalakim*, equivalent to 80 minutes, which breaks down into 76 *regaim*, very fractionally longer than the modern second.

All these ancient units functioned only in the realms of astron-omy, astrology and mathematics. Small units of time were meas-ured by the people in far more practical, if less exact ways. This was the case right up until the end of the sixteenth century. In non-scientific literature, writers were left to their own devices when it came to describing durations of seconds and minutes. For instance, an earth tremor of 23 August 1295 is described in the *Chronicle of Constance* as 'a mighty quake which occurred around noon and lasted for as long as it takes to say a Paternoster and an Ave Maria'. Not a particularly accurate way of telling the time. If you say the two prayers at a reverential, normal Sunday-morning pace, you can expect to keep going for about 35 seconds. However, in the garbled express-train manner of the Tridentine Mass, you might, if your Latin is good enough, be finished in under 15 seconds. Either way, it seems likely that the *Chronicle of Constance* exaggerated the length of the quake. Coincidentally, another earthquake nearly 200 years later was recorded in much the same way. In a letter, the Neapolitan quake of 1456 is described as 'lasting the time it would take to say the Miserere quite slowly and, more specifically, one and a half times'. We might conclude that the author of this passage, a Signore Paolo Rucellai, actually said his prayers at the moment of the quake, which is how he managed to be so precise. This is more likely than the idea that events were frequently measured according to the time it took to say particular prayers. The Paternoster (Lord's Prayer) and the Ave Maria (Hail, Mary!) are, after all, both prayers of absolution. 'Pray for us sinners now and at the hour of our death' goes the Ave Maria while the Miserere, which comes from the Latin Mass and forms an important part of the Requiem, is also something that a committed Christian might well start spouting as the earth begins to shake beneath his feet.

Curiously, these and other prayers crop up several centuries later as official time units to control the use of torture. In 1532, Charles v, Holy Roman Emperor, decreed that torture victims should not suffer beyond a certain limit. At first it was suggested that either the Ave Maria, the Paternoster or the Miserere should be read aloud to provide the maximum length for which it was permitted to suspend a victim by the arms tied behind his back. This obviously left it still wide open for the torturer to decide how fast or how slowly he wished to say the prayer. He could span it out for several hours, if he so felt, intoning just one word every 40 minutes.

The inherent uncertainty of this rule led Pope Paul III ('the Pope with a permanently itchy nose') to impose a limit of one hour on torture, which was to be upheld with the use of a sand-glass. This may have been a fairer system, especially if the victim was allowed to keep an eye on the sand-glass himself. Unfortunately, it was noticed that victims of torture, by fixing their gaze on the sand-glass, were able to ride the roughest tortures in the belief that it would all be over as soon as the last dribble of crushed eggshell dropped down into the bottom glass. It gave them strength, which was so counterproductive to the purposes of torture that in the end it was decided that sand-glasses must be kept out of the sight of torturees. In the late eighteenth century the Austrian criminal constitution of Empress Maria Theresa banned the practice of warning torture victims what they were about to receive and how long it would last, worried that the system would break down for 'strong people and obstinate Jews accustomed to denying things'. We digress.

Minutes themselves, as we have seen, may have been recognised as early as 1345 and frequently referred to in the annals and *computus* of ecclesiastical literature, but that is not the same as stating that for normal, lay Europeans minutes served any practical purpose until at least 250 years later, as Gerhard Dohrn van Rossum, a historian of the hour, passionately points out:

> Misunderstandings in historical literature arise when one fails to realise that up until the end of the sixteenth century, the clock-hour was the 24th part of the full day or the twelfth part of the day, or the night, at the two equinoxes. In everyday life this hour was divided into halves, thirds, quarters, sometimes into twelve parts, but it was not divided by sixty or understood as the period of sixty minutes.

If this is so, then what were people using to describe small units of time? There seems to be a huge and varied number of ways which have been used to articulate such fleeting moments.

'He ran into the woode and over Greenston's hedge as faste as rabbit vereteth,' says the old tale of Abbé Wishart. If the thirteenth-century rake had ever heard a rabbit fart, we cannot be sure, but the point he is no doubt trying to make is that his protagonist ran into the wood and jumped the hedge extremely fast. Unless of course rabbits' farts are notably slow affairs – we do not know. Traditionally, the fart is used to denote the coming of a new day or a new season, and there are very few examples, like Abbé Wishart's, where it is used for a measure of duration. In vulgar parlance, even today the phrase 'I was up at squirrel's fart' is used to mean 'I got up extremely early'. *The Oxford Book of English Verse*, for generations the staple classic anthology, has always, in all its many and varied editions, started with an intriguing and anonymous poem called 'Cuckoo Song':

> Summer is y-comen in,
>   Loude sing, cuckoo!
> Groweth seed and bloweth meed
>   And sprin'th the woode now—
>   Sing cuckoo!
>
> Ewe bleateth after lamb,
>   Low'th after calfe cow;
> Bullock starteth, bucke farteth.
>   Merry sing, cuckoo

In rural Somerset time was measured off against a walking-stick. A measurement would be shown by marking the thumbnail against a knobbly walking-stick at different distances in relation to the end of the stick. This practice was common among the rural communities of western Somerset right up until the mid-1970s but is sadly all but forgotten now.

In China in the eighteenth century, minutes were counted as

'rice-bowls', for it was deemed rude to eat a bowl of rice either too fast or too slowly. What the correct rice-bowl eating time should be has not been easy to find out. Let us assume that it was around about five minutes, for in Lee Chiang's charming children's fable, first published in the *Golden Dragon Collection* of 1879, the young boy Xue Wei is asked to run from his parents' house to the doctor's to inform him of a strange beast which has visited the house every night for seven days and has now bitten off grandmother's foot and disappeared into the forest with the limb between its terrible jaws:

> Little Wei ran as fast as he could, he ran and he ran but there were 40 houses between him and the doctor, the sand was deep and the wind blew hard in the young boy's face. 'I could run to the doctor's house in the time of three rice-bowls without this evil wind,' Wei said aloud to himself, but just at that moment a golden eagle swooped down from the high cloud and picked up the boy, with his sharp talons gripping on to Wei's thin arms. They flew so fast that Wei could not see, for the biting wind was turning his eyes into salt. He started to cry out, but in the time of only one rice-bowl the eagle had delivered him safely at the doctor's door.

Whether this means that the Chinese eat their rice very slowly or that Chinese eagles fly very fast is far from certain.

In England, from the sixteenth century the use of 'minutes' in speech denoted a vague period of time, usually an instant or short moment. In Shakespeare's *A Midsummer Night's Dream*, for instance, we hear the cruel Lysander say: 'Content with Hermia? No, I do repent the tedious minutes I with her have spent.' Or, as in his fourteenth sonnet: 'Nor can I fortune to brief minutes tell, Pointing to each this thunder, rain and wind.' In Lyndesay's *Monarche* of 1552: 'The small minuth of one hour to them shall be so great dolour, they think they shall have done remane an thousand yeir in to that pane.' In these examples the minute is clearly used as a figure of speech. It is not until the eighteenth and early-nineteenth centuries that examples of minutes, as we define them today, find their way into the common language. A clear example of this is in Thomas Love Peacock's prose satire *Nightmare Abbey* of 1818, where we are told: 'The hour-hand passed the VII – the minute-hand moved on; – it was within three minutes of the appointed time.'

And so, for the dutiful definition of a minute as it now is: like many units of time the minute was standardised in 1967 and is now represented as a meaninglessly large number of atomic oscillations in the cesium atom. Bede would have had no truck with this; the former definition of a minute as $\frac{1}{60}$ part of an hour or $\frac{1}{1,440}$ part of a mean solar day would have been far more palatable to the holy man from Jarrow.

CHAPTER IV

HORAE
Hours

W hy do we have 24 hours in a day? Perhaps we can blame the Egyptians for that, because it was in settlements along the Nile, more than 3,000 years ago, that the holy priests of Ra ordained that there should be 12 hours of daytime and 12 hours of night. The Sumerians, as we have already seen, were the first to give prominence to the number 12, for the basic reason, we suspect, that they, like us, had 12 phalanges on their fingers and that these handy bones proved immeasurably useful for counting. From this innocent, market-led beginning, the number 12 grew into something mystical: a powerful symbol that represented gods and planets and divided neatly into other numbers that represented even more gods and even more planets.

For this reason the Egyptian year was made of 12 equal months of 30 days each. The total number of days was made to correspond to that wholesome, ordered number, that pure representation of cosmos, 360. That the solar year was in point of fact 365 days long was known to the Egyptians, but it worried them not. For a year, to an Egyptian, was 360 days, that is all there was to it.

The great falcon-headed god Ra, ruler of the sun, would punt his boat across the sky every day and at night-time would sail another vessel on a tour of the underworld. Between dusk and dawn the Egyptians could sleep tight in their beds secure in the knowledge that Ra would be smashing the evil serpent Apepi on the head with his hammer and that when Apepi was vanquished, as he invariably was every single night, the great Ra would allow daybreak to dawn. He would rise from the ocean of chaos on a primeval hill and recreate himself as Ra so that he was reborn to the people of Egypt every day. And the first thing he would do as he popped his ornithological head over the horizon would be to create eight more gods. By the time that this elaborate religious tapestry had sunk into the minds of the Egyptian people, Sumerian phalanges were all but forgotten. For the Egyptians, 360 was the sacred number. It was the numeral that Ra had chosen and specially ordained for the number of days in the year. To compensate for the difference between the 360-day calendar year and the 365-day solar year, Ra, in his infinite kindness, gave the people of Egypt five epagomenal days in which they were permitted to worship him and all the other gods. But at no point would an ancient Egyptian accept the assertion that he was only interested in 360 because Sumerian fingers had been made that way. Such a suggestion would be blasphemy, and the punishment for that was slow, ritual death.

That the days should be divided into 12 equal hours was an obvious, easy decision for the Egyptians to reach. Don't forget, they adored the number 12! The problem they had was matching equal hours to days and nights of unequal length. Eventually, they discovered that days and nights were only of equal length at the two equinoxes, and so the 12 hours of daylight and the 12 hours of night became merged as a variable 24-hour day. We seem to have vacillated back and forth on this issue. Early clocks all tolled 24 hours, but people lost count of the *boings* and complained. After this the clock reverted to two periods of 12 hours each, marked as a.m. (*ante meridiem*) and p.m. (*post meridiem*), signifying time before and after noon. Later still, in the twentieth century, armies and timetables and others with exacting standards reverted back to the 24-hour clock, aware that a misunderstanding of the a.m./p.m. system might, one day, lead to catastrophe.

A brief digression is needed to explain the equinoxes. As the earth orbits the sun it is rotating on its axis. At each end of the axis are the poles, the North and South Poles. The earth is tilted so that, most of the time, one of the poles is closer to the sun than the other. If we imagine that the North Pole is leaning away from the sun and the South Pole is leaning towards it, and as the earth orbits the sun it remains at its tilted angle, then when the earth reaches the far side of the sun the North Pole will now be leaning towards the sun and the South Pole will be leaning away from it. Try to imagine that whole sequence again.

The North Pole is leaning towards the sun. This means that it is light, and as the earth rotates it remains light because no part of the pole is obscured by shadow. The South Pole, by contrast, is in permanent darkness, for, however the earth rotates on its axis, the pole is always leaning away from the sun, thereby not getting any light. It is at this time that the countries of the northern hemisphere have their summer; for when the North Pole is leaning towards the sun the days are longer and warmer. As the earth passes to the other side of the sun, the tilted angle remains the same but now it is the South Pole which is leaning towards the sun and so it is the countries of the southern hemisphere which are most exposed to the light and which are now experiencing long, warm days. Meanwhile, the North Pole, which is tilting away from the sun, is in darkness. The closer you are to the North Pole (Iceland and Greenland, for instance), the shorter your winter days will be. As the world continues on its orbit, there will come a time when the poles are equidistant in relation to the sun, and at that point the night and the day are of equal length. That is an equinox. Each year has two equinoxes. After one, the days draw in and the nights get longer; after the other the days get longer and the nights get shorter. The longest day and the longest night (i.e. the point at which a pole is at its closest or furthest from the sun) are called respectively the summer solstice and the winter solstice.

So that explains what equinoxes and solstices are, but how did ancient and primitive people manage to identify them? The important thing is to determine the fixed point that divides the day into two: namely, noon, the moment at which the sun is at its highest. This can be done only with two fixed points, one, a point on the horizon, and the other a nearer object, say a stone, or a window frame. It is for this reason that the annual observation of the equinoxes and solstices can have been possible only since people have lived in fixed abodes. The early Cenozoic era, when *Homo erectus* was wandering around scavenging, fighting and eating strange uncooked food, would not have yielded any intelligence on this subject.

And, as a small extension to this sprawling *divertimento*, we come to the issue of daylight-saving. Putting the clocks back or forwards is, in many ways, a synthetic gesture, since we cannot really *save* daylight hours by turning the clock back or forwards, the point is simply to make us wake up an hour earlier or later so that we enjoy as much of the sunlight as possible. If we want to enjoy an extra hour of daylight in the summer, we move the clock hands forward by one hour. This means that the sun will go down at eight o'clock instead of seven o'clock, as it did on the previous evening. It also means that sunrise, which was at 5.30 a.m. is now at 6.30 a.m. This is all very well for humans, but farm animals wish to be fed at the same hour, irrespective of what time it is by the human clock. Animals fed at 6 a.m. as the sun is rising have to eat their meals in the dark during British Summer Time.

Back to the track. While the Egyptians were determined to see their days and their nights as containing 12 hours each, they faced a problem. How could they measure hours throughout the day and throughout the night when they had very different ways of counting each? How could they be sure that the hours they measured on their sundials in the day were the same as the hours they measured by astronomical observation at night?

For measuring daylight hours they devised very simple shadow-clocks made with two pieces of wood. One was flat, like a ruler, and the other, a shorter block, crossed the first at 90 degrees at one end. By laying the shadow-clock down so that the length of the ruler was running from east to west with the top block to the east, the sun would cast a shadow of the top block along the length of the ruler. As the sun slowly rose in the east, the shadow would shorten until, at midday, the sun would be roughly overhead and the shadow at its shortest. At this point the Egyptian time-measurer (or, more likely, his servant) would turn the shadow-clock around 180 degrees so that the rest of the day could likewise be measured by the increasing length of the shadow on the ruler as the sun set in the west.

Shadow-clocks were prevalent all over the ancient world. Menhirs, megalithic stones, Egyptian obelisks and simple sticks have been used as a gnomon (the pin of a dial whose shadow points to the hour, gnomon is also an obsolete and jocular name for a nose) to tell the time by the varying lengths of their daytime shadows. By the first century AD, sundials were all the rage in cities around the Mediterranean, where they were fixed to public buildings, baths, temples and private houses. There was even a portable travelling version which could be adjusted to take into account unequal day lengths at different degrees of latitude.

Timekeeping at night was more difficult and required a certain degree of astronomical knowledge to make it work. In Egypt they watched the stars as they rose in the sky as the sun was setting. Unsurprisingly, they found 36 stars that were particularly effective. These were known as the *decans*. Each night they would cross the sky from east to west, marking off the hours. From the thirteenth century, night-time hours could be measured with a device called a nocturnal. This converted the movement of the stars into something resembling a modern clock-face. It had dials that could be adjusted to take account of the date. By holding it up at arm's length you could squint through a hole at the Pole Star while adjusting the dials to align with stars from the constellation of the Plough. As the hours move on, the Pole Star will be seen to rotate its position in relation to the Plough, allowing the nocturnal to give a simple and reasonably accurate account of the time.

Plotting the time by the stars was older in fact than the lunar-solar observations which brought about the calendar. It was common for all educated Romans and Greeks to have a general-knowledge grounding in the stars, which was far better than ours is now. They could all recognise the major constellations and point to over 40 stars by name. Many stars to the Greeks and the Romans had a significance far beyond the distant twinkle of light in the far night sky, as Homer shows in this famous passage from the *Iliad*, written nearly 3,000 years ago:

> Him first Priam saw with his old eyes,
> And o'er the plain he lightened, dazzling bright,
> Like to the star that doth in autumn rise,
> Whose radiant beams, pre-eminent to sight,
> Shine with their fellow stars at noon of night:
> Orion's Dog we mortals call his name:
> Sign is it of much ill, though clear its light,
> And mighty fever brings to man's poor frame:
> So, as he ran, the brass upon his breast did flame . . .

We mortals nowadays use the name Sirius, but to Homer he was Orion's Dog, the evil star whose rising coincided with the hottest period of summer, when fever was rife. All problems that took place at the time of the year in which Sirius was rising were blamed on that star. The watchman who speaks the prologue at the beginning of the *Agamemnon* by Aeschylus uses the stars both as a clock to measure the night hours and also to foretell the seasons. He is also of the opinion (500 years BC), as many ancients were at that time, that the stars were the cause for everything, both good and evil, which happened in this world. Shakespeare allows Cassius in *Julius Caesar* to take a radically un-Roman line on this: 'Men at some time are masters of their fates: The fault, dear Brutus, is not in our stars, But in ourselves, that we are underlings.' Not to Aeschylus's watchman:

> On elbow bent, watching, as 'twere a dog,
> I mark the stars in nightly conclave meet.
> And those bright constellations, without peer,

> Lord paramount in heaven, that winter bring
> And summer in their train for mortal men,
> Right well I know them as they come and go.

A system that relied on nightwatchmen after dark and shadow-gazers in the daytime was bound not to provide the whole answer to our human timekeeping needs.

Water-clocks, of which the oldest is said to date from 1500 BC, should, in theory, solve the problem of daylight and night-time reckoning. Why should water care if it is night or day? 'Ol' Man Ribber jus' keeps rollin' along', as they say in *Showboat*. Unfortunately, water-clocks, or clepsydras as they were known to the Greeks (literally 'water-thieves'), were highly inefficient and they *did* care whether it was night or day. Most water-clocks were based on a simple premise. They may have become extremely complicated as dissatisfied customers worked out ever more ingenious ways to regulate them, but however many floats and bobs, dams and paddles were fixed inside them, the basic idea remained the same. Water drips out of one vessel into another. The vessels are marked and the markers tell the time as the level rises in the bottom vessel or drops in the top one. If a water-clock were kept outside, the difference in temperature between night and day, which in Egypt could fluctuate between 25°C and 1°C, would be enough to double the flow of water from the top to the bottom vessel at night-time. Clepsydras were put to practical use by the Roman judiciary. Court cases were supposed to last no longer than a day, and so each party, the prosecution, the defence and the judges were allowed an equal amount of time to put their case. They were allotted buckets of water, not to throw over each other but to pour into the clock, allowing each of them to speak for as long as it took to drain the water from the top to the bottom vessel.

At the twilight of the Sung dynasty (*c* 1100) the Chinese emperor, Sung Che-tsung, erected an enormous water-clock at an inn in the capital, Pien. It was an extraordinary affair, a building 40 feet (12 metres) high and 20 feet (6 metres) wide which served as a clock, an observatory and a centre for scholarship. Water gushed in from a river source, filling buckets that were each tied to the circumference of a 12-foot (4-metre) revolving wheel. When each was full, the weight of water would cause the wheel to rotate to a position where the next bucket would fill and the previous one would empty its load. From the main wheel a Heath Robinson chain of cogs and belts would drive a rotating disc with puppets on it. Each puppet would leer with its haunting white face from a window in the clock. The time could be deduced by whichever puppet happened to be showing its face. Similar huge clepsydra were built in Greece 1,000 years earlier. They were all eccentric devices, and most of them did not tell the time very well. In about 50 BC the citizens of Athens pooled together to erect a giant clock in the market-place. The result of this was the Tower of the Winds, designed by Andronicus of Kyrrhos, which still survives, in ruined form, today.

We have heard how Alfred the Great used candles. Similarly in Sparta, they used 'smell' clocks, which burned a series of incenses, each denoting a different hour of the day.

More common than joss-stick clocks were sand-glasses, used nowadays for timing boiled eggs. These elegantly simple devices allow sand to sift though a narrow aperture between two glass containers. Ironically, they are filled not with sand, which would erode the glass, but crushed eggshell, which is softer and breaks down into smaller particles.

Before 1500 AD there was only one place in western Europe where the day was regularly and systematically ordered, and that was in the Church. For monks in a medieval monastery, the day was marked out by a strict succession of prayers, and these, seven in all, became known as the seven canonical hours.

'Doubt not, worthy senators,' says Milton in the *Doctrine of Divorce*, 'to vindicate the sacred honour and judgement of Moses your predecessor, from the shallow commenting of scholastics and canonists.' He is talking about the people who lay down the rules of the Church and stringently follow them. Throughout the

Middle Ages, these dedicated men trawled the scriptures looking for clues as to how God intended his flock to behave and then decreed canonical laws for everyone to obey. They were the Church's equivalent of a modern local authority's safety regulations officer – which might go some way to explaining why Milton held them in such low esteem. But this is how they made their laws. St Luke tells us that the Psalms (Hebrew sacred poems) should be a source of guidance to followers of Christ, and St Paul demands that we 'sing psalms and hymns and spiritual songs'. From here, the scholastics and canonists trudge through the Psalms, all five books of them, looking for evidence of God's intention. David, the sweet psalmist of Israel in Psalm 119 says:

> Princes have persecuted me without a cause, but my heart standeth in awe of thy word. I rejoice at thy word, as one that findeth great spoil. I hate and abhor lying: but thy law do I love. Seven times a day do I praise thee because of thy righteous judgements.

And so it was decreed that canonical hours should be added to the rigid, monastic timetable. It started at sunrise with Matins and ended with Compline at about 8 or 9 p.m., and between them there was praying at Prime, Tierce, Sext, Nones and Vespers. Perhaps David, in his enthusiasm for the number seven, forgot that on the seventh day the Lord rested – such are the inconsistencies of religion.

William Hazlitt, the nineteenth-century essayist, had little to say for the canonical hours and took a cynical view of the whole system of time-measurement in his sketch *On a Sundial* of 1839:

> Most of the methods of measuring the lapse of time have, I believe, been the contrivance of monks and religious recluses, who, finding time hang heavy on their hands, were at some pains to see how they got rid of it.

* * *

The *Book of Hours* began to appear in the thirteenth century. It contained bundles of prayers that were intended to be read at each of the canonical hours. How much attention was ever paid to these strictures is not easy to tell. The books were widely printed, but this does not tell us how often they were used. The Bible, after all, is the best-selling book in the history of publishing, but the people who have read it from cover to cover are few and far between. The inconvenience of having to say prayers at seven different intervals throughout the day can be imagined without much difficulty; the boredom of such an undertaking defies all common sense, but the *Book of Hours* survived its function and became, during the fourteenth and fifteenth centuries, the most popular book in Europe. Private libraries were not universal, even among the rich, until the sixteenth and seventeenth centuries. People did not own many books. The *Book of Hours* was one of the first 'must-have' volumes in the history of private book-buying, and if people did not use it in the manner for which it was prescribed, that hardly seemed to matter, for here was a chance to show off. He who possessed the handsomest copy of the *Book of Hours* might consider himself the most worthy, the most pious, the most learned and the most powerful of all. And so the European aristocratic and moneyed classes set about commissioning the most extravagant copies imaginable. Patronage for these books was crucial to the development of Gothic illumination. Some of the grander versions took many years and many people to paint, assemble and bind them. Of the most spectacular, the Duc du Berry's *Très Riches Heures*, now held at the Musée Condé at Chantilly, is bound with glittering jewels and contains dazzling colour pictures showing, in extraordinary detail, the Duke's palatial homes, episodes from the lives of the saints and biblical stories, and a magnificent set of illuminated calendar pages. It was executed by Pol de Limbourg and his five brothers between 1414 and 1418. The *Book of Hours of Charles d'Angoulême* and *Livres d'Heures de Rohan*, fourteenth-century examples, both at the Bibliothèque Nationale de Paris, provide further evidence of the wildly sumptuous spirit in which the *Book of Hours* was regarded. 'Blessed are the meek,' said Jesus, 'for they shall inherit the earth.' Christianity is full of these strange paradoxes, but nothing in the Bible would dissuade these vain-glorious men from dressing their books in royal robes. It was a holy book, after all. 'Lay not up for yourselves treasures upon earth, where moth and rust doth corrupt, and where thieves break through and steal' – and then, it should be added, 'unless, of course, they be encrusted upon ye mighty *Book of Hours*.'

Prime, Tierce, Sext and Nones continued to be used by the lay community to signify quarters of the day, both before and after the widespread use of public clocks. Town clocks, right up until the nineteenth century, chimed quarters in preference to hours.

In the Middle Ages night-time hours were also given names, even though these names were of little or no use to the general public: *occasus solis* was sunset; *crepusculum* the hour of dusk; *vesperum* marks the appearance of the evening star in the west; *conticinium* means silence; *intempestum*, when all activity is ceased; *galicinium*, that infernal hour when the cock crows; *aurora*, first light; *diluculum*, dawn; and *exortus solis*, sunrise. These hours were not of equal length and, in all probability, meant very different things to very different people.

From the end of the thirteenth to the beginning of the four-teenth centuries, the churches and the royal palaces were gradu-ally set up with their own clocks. It was at this time that the canonical hours became absorbed into the 12-hour system. There was no clock in the world that could strike *dusk* or *dawn* and, as we have seen, the 24-hour day containing 1,440 minutes and 86,400 seconds was established by consensus among the monks of the Christian Church during the early Middle Ages. It was only natural therefore that clocks, when they came to be made, should be designed to the 12-hour system that had for centuries been acknowledged, if not actually used, by the Church. There was one small problem, however; once 12-hour clocks had established themselves as a common feature of the churches and palaces all over medieval Europe, it became immediately apparent that they were all chiming away and telling different times in different places. Time needed to be synchronised across the land. Charles le Sage, the 'wise' king of France (1338–80), took it on himself to sort it out. Charles was a man of luxury. He surrounded himself with intelligent people, who shared his own intelligent points of view. In a statue by an unknown sculptor in the Louvre, Charles looks wet, but sympathetic, with a small chin, hair flopping below his jaw, a measly penitent gaze, and a crown uncomfortably screwed to his head. But in life, Charles was no wet. He was a brilliant military commander who won so many victories against the English that the dreadful settlement of 1360 (in which his compatriots had handed the British most of south-western France and three million gold crowns as ransom for Charles's father, Jean le Bon) was practically nullified by the end of his reign. He successfully reorganised the army, the navy, the French tax system, and at the same time made important alliances with Portugal, Spain and Flanders.

Among Charles's many interests were books (he instituted a magnificent library at Vincennes) and clocks. For le Sage, the day divided itself between hours of work and hours of prayer, which hours he counted by wax candle-clocks that were kept burning in several of his palace chambers at a time. He also had mechanical clocks installed in the palaces at Vincennes, St-Pol and the Louvre. His important contribution to the history of timekeeping stems from a proclamation of 1375 ordering the churches of Paris to chime their bells according to his clocks and not according to their own whims and fancies. Critics have suggested that, as a bookish, prayer-obsessed man, Charles might simply have been irritated by the churches of Paris ding-dinging all over the city at all hours of the day and night. Others have suggested that Charles was more interested in diluting the power of the Church than he was in synchronising the use of clock-time throughout his land. Whatever his motivation, by issuing this decree Charles v of France put another foot into the history books as the father, perhaps, of modern time-reckoning. (Some historians are of the view that Charles's decree was legendary and not actual, hence 'perhaps'.)

In 1924 the BBC broadcast six pips over the radio for the

British nation to synchronise its watches on the hour. It may not seem like an important, or even a particularly interesting event now, but in 1924 it must have seemed as momentous as Charles V's decree did to the priests in 1375. According to Jonathan Betts, curator of horology at the Royal Observatory, Greenwich: 'It was regarded as one of the wonders of the age. From that moment on, everyone had time in their home. It was seen as a great thing.' The BBC was attempting to synchronise the accurate watches at Greenwich with the less accurate watches of His Majesty's subjects. In the early days of the BBC, radio announcers, who dressed in dinner jackets for work, used to attempt to mimic the chimes of London's most famous clock, Big Ben, by playing notes and chords on a studio piano. Later, tubular bells were introduced. New Year 1924 was ushered in with a live broadcast of the Big Ben bongs. Thereafter, the pips were broadcast around the world, first from Greenwich, where they were regulated to Frank Hope-Jones's free pendulum clock, and in 1939 from the magnetic observatory at Abinger in Surrey and then, until 1990, from the beautiful ancient village of Herstmonceux in Sussex. By 1972 the famous six short pips had turned into five short and one long. At one o'clock in the afternoon of 4 June 1990, the BBC broadcast, for the last time down a telephone line, its now world-famous pips.

'Up until that time,' explained Mr Betts at a party to celebrate 75 years of BBC pipping, 'we had been fudging timekeeping somewhat. We had to stretch and squeeze seconds to take account of the variations in the speed of the earth's rotation.' In 1967 atomic-based, coordinated Universal Time replaced Greenwich Mean Time as the world's official standard. Nowadays the BBC broadcasts its pips from atomic clocks linked by satellite to a dingy radio transmitter in Rugby.

CHAPTER V

DIES
Days

W ho was Solomon Grundy? We seem to know a bit about him:

> Solomon Grundy,
> Born on Monday,
> Christened on Tuesday,
> Married on Wednesday,
> Took ill on Thursday,
> Worse on Friday,
> Died on Saturday,
> Buried on Sunday.
> This is the end
> Of Solomon Grundy.

Was he someone whose name just happened to rhyme with Monday? Or was he a real person perhaps? The Solomon Grundy nursery rhyme was first printed in James Orchard Halliwell's *Nursery Rhymes of England* of 1842. Where Halliwell got it from, no one knows. A similar rhyme, attributed to 'a facetious Jacobite, on the occasion of the early death of some infant princess of the reigning house', appeared in the *Bristol Observer* in 1881:

Little Goody Tidy
Was born on a Friday,
Was christened on a Saturday,
Ate roast beef on Sunday,
Was very well on Monday,
Was taken ill on Tuesday,
Sent for the doctor on Wednesday,
Died on Thursday.
So there's an end to little Goody Tidy.

Keen-nosed critics might spot the difference. Whereas 'Little Goody Tidy' is a fine example of harmless British whimsy, 'Solomon Grundy' seems to be telling us something. It is an altogether more baleful effort. Both Goody and Grundy die, it is true, but the events of Grundy's life are more significant, more resonant, even, than the events of Goody's life. So Goody is clearly a parody of Grundy, with a little bit of 'This little piggy went to market' thrown in for good measure, and is therefore of no real consequence. The same, you may well think, is the case for Grundy – unless, of course, you happen to be interested in the days of the week.

It is a strange fact that, while all the names for our months derive from ancient Rome, the days of our week come from the Anglo-Saxon language. It is well known that the seven-day week has been in use for over 3,000 years, but there are a number of coincidences between the Anglo-Saxon names of days and the Romans names which raise the interesting question about the Romans and the Anglo-Saxons: who was influencing whom?

The Romans, obsessed with their gods, named not only their planets but their days of the week, as well as some of their months, after them. Thus, Mars, the god of war, Mercurius, god of trade, Jupiter, god of power, Venus, goddess of fertility, and Saturn, who represented agriculture, among other things, were all given planets and days of the week in their honour. At the same time the peoples of northern Europe, the Anglo-Saxons and the Teutons, also had their gods, which were also associated with planets and (you've guessed it!) days of the week. The Teutonic god of war is Tiw or Tyr. His name lends itself to Tuesday (Tisdag in Swedish). In French, Tuesday is *mardi* and it is *martedi* in Italian, obviously names based on the Latin *Dies Martis* and referring to the Roman war god Mars. Similar 'coincidences' crop up across the board. Freya or Frigg is the Teutonic goddess of fertility; Venus is the Roman equivalent, and from these two names we derive 'Friday' on the one hand, and *vendredi* or *Dies Veneris* on the other. Thursday, the fifth day of the week, is the same: in Latin it is *Dies Iovis*; *giovedi* in Italian and *jeudi* in French, in all cases named after the Roman god of power, Jupiter. The Teutonic equivalent is Thor or Thunar, wherein lies the root of our word 'Thursday'. One could go on, but the point is surely made. Anglo-Saxon names of the days seem to be translations of the Roman names – or is it possible that the Romans were themselves influenced by people and the cultures which they found at the northernmost corner of their empire?

Before we set out on that investigation, a brief return to Solomon Grundy. We think we know next to nothing about him, and that he might even be a figment of some whimsical rhymster's imagination. But one question arises. If his name was Grundy only because it proved to have a useful rhyme with Monday, how come he wasn't born on Sunday which rhymes just as well and has the added advantage of being the first day of the week?

A close examination of the text reveals that it is not just a trifle but a key to some of our oldest myths about days of the week. Monday takes its name from the moon (as does 'month'), and in English folklore a child born under a full moon will be handsome or beautiful (depending on sex). Curiously, the same myth pervades many Norse legends, with the difference that beauty was bestowed on babies born under a new moon, not a full moon. The point is echoed by another days-of-the-week nursery rhyme which starts: 'Monday's child is fair of face.' Grundy's Wednesday marriage is also significant. It was considered a good day on which to marry. This tells us that Grundy was not only traditional in his ways but superstitious in his views – a cautious, handsome, god-fearing man, so far.

Nowadays, Christian weddings usually take place on a Saturday. This is purely a matter of convenience. Weddings are bigger than they once were, and the holy sacrament of marriage forms a smaller part of the total proceeding. Weddings are now about marquees; one, two, possibly even three meals; dancing, confetti, horrid music, hundreds of guests and offensive best-man speeches. To ensure the attendance of guests (often arriving from far corners of the globe) modern weddings are set for weekends, and since priests are invariably busy on Sundays, for purely practical reasons, the wedding has become a Saturday 'do'.

All of this is recent history. We mustn't forget that Solomon Grundy was dead by 1842, and he probably lived a long time before that. In those days, weddings were comparatively modest affairs that seldom took place on a Saturday. Wednesday was considered particularly propitious; it was a lucky 'red-letter' day. In the Book of Genesis, God created the moon, 'the lesser light to rule the night', on the fourth day, 'and God saw that it was good'. The Persians, from whose ancient ancestors the Book of Genesis is deemed to have derived, maintain, for the same reason, that the fourth day is a lucky day.

Thursday is the day of Thor, the great Norse god with a crashing hammer called Mjollnir, who failed to smash the skull of the serpent Jormungand, symbol of evil. Thor is the god of thunder, too. It is apt that Grundy should have been struck down on Thor's day.

That he got worse on Friday is hardly surprising either, for it has long been held that Friday is an odious day and one that brings nothing but ill-fortune. Friday the thirteenth is supposed to be even worse for 13 is an unlucky number. At the Last Supper, a luncheon attended by 13 men, Jesus, between bites of fish, revealed that someone around the table would betray him.

The superstitious Romans called Friday a *Dies infaustus* ('unlucky day') because it was the day on which their mighty armies got their asses kicked at the Battle of Gallia Narbonensis. Friday is also treated with suspicion by the Buddhists and Brahmins. Christians, too, are wary of Fridays, avoiding them whenever possible for marriages, christenings or other ceremonies that require propitious omens. This may have been inherited in part from the Romans, but is more obviously connected to the crucifixion of Jesus that took place on Good Friday (which is not to suggest that Grundy is a religious symbol). Incidentally, in case you think you have spotted a contradiction here, the term 'Good Friday' does not invoke the usual sense ('to be desired' or 'approved of'); it simply means 'holy Friday'. Some Christians believe that Adam and Eve, who caused the downfall of our entire species by eating the forbidden apple, committed their atrocity on a Friday. There are many other curious phrases in the English language which testify to our long-held mistrust of this day. Geoffrey Chaucer, in *The Nun's Priest's Tale*, a parable about a cock and a fox which warns against idle boasting, ascribes the animals' misfortune to the day of the week: 'And on a Friday fil al this meschaunce,' he says. In America it is still averred that a 'Friday moon brings foul weather', and nail-cutters should take heed: 'Cut them on Friday, you cut them for sorrow', or, more bluntly, 'A man had better ne'er be born As have his nails on a Friday shorn'.

Saturday was the day of Grundy's death; the lead that lines his coffin is the lead of Saturn. It is an evil, gloomy, heavy-hearted planet, and bad luck is said to fall on those who are born under its glower, as Pholomeus reminded us: 'The children of the sayd Saturne shall be great jangeleres and chyders and they will never forgyve tyll they be revenged on theyr quarell.' The French have a terrifying figure called Baron Samedi (Lord Saturday) who is a skeleton dressed in a black frock-coat as a Victorian undertaker. Is he the shadow of Saturn?

In Keats' epic poem *Hyperion*, written in 1819, the glowering Saturn is given a leading role:

Deep in the shady sadness of a vale
Far sunken from the healthy breath of morn,
Far from the fiery noon, and eve's one star,
Sat grey-haired Saturn, quiet as a stone.

Grundy's death occurring as it did on a Saturday was therefore indicative. So, too, was his burial on the very next day. Sunday in the Romance languages is *domenica* (Italian), *domingo* (Spanish) and *dimanche* (French). All of these mean 'the Lord's day', the day that Christ ascended into heaven, for which reason good Christians are expected to attend church on that day. The English name 'Sunday' derives quite obviously from the star, and the rising of the sun is a frequently used metaphor for Christ's ascension. A metaphor for the setting sun might be applied, with equal force, to the burial of Solomon Grundy. That he was buried on a Sunday, the Lord's day, is sufficient evidence to conclude that he was a God-fearing Christian man, probably heading for heaven. So here he is, Solomon Grundy, fleshed out as never before: his parents were Christians, as was Grundy himself. He was a handsome (though short), married, superstitious, conventional, God-fearing, peaceful, unfortunate man.

Just a minute: peaceful? Short? Where did they come from? A bit of a cheat these, but the name Solomon derives from the Hebrew *Shalom* meaning 'peace', and 'Grundy', according to *The Oxford English Dictionary*, is a designation applied to a small person. The dictionary offers a citation from 1570, being a description of a John Vander Warfe of Antwerp: 'Of some he was called Shildpad for that he beyng a short grundy and of little stature, did ryde commonly with a great broad hat.'

So Grundy was short – either that or, according to the same dictionary, he was 'granulated pig-iron' which cannot be right!

So it seems that the names of the days derive from the Roman astrological week, which is not connected to the Jewish seven-day week, except by association with the Babylonians. The Romans, as we have seen, ascribed successive hours in the day to the seven known planets, naming each day (of 24 hours) after the planet that was supposed to rule its first hour. Thus, the Roman names spread throughout their empire and were adopted, in translated form, by the Britons, even before Caesar's invasion of 57 BC. The German-speaking Teutons took the Roman names and changed them to the Teutonic names with which they were supposed to correspond.

During the French Revolution, Philippe-François Nazaire Fabre d'Eglantine (about whom we shall have plenty to say in due course) was put in charge of renaming the days on behalf of the French people. Fresh from chopping the heads off half the aristocracy, the revolutionary forces were looking to cleanse their calendar of all ancient and ecclesiastical connotation. Eglantine, a Romantic poet of sorts, quickly drew up a list of names for the days. Out of the window went *lundi*, *mardi*, *mercredi* . . . in flew *saffran*, *raisin* and *pomme*. This was not all. The system they devised was a 12-month year with each month consisting of three weeks (decades) of 10 days each. The names of the days in each decade differed in each month, which meant that the calendar had 360 days, each with a different name (10 times three times 12 names). The only rules Eglantine applied were to name the fifth days (known as 'quintitis') of each decade after animals, the tenth days ('decadis') after farm implements, and the remaining 288 days were named after minerals, vegetables, fruit or flowers. Days were thus named: dog, pig, chicken, donkey, rabbit, mullet, bumblebee, rose, tomato, zinc, plough, aubergine, pepper, lilac, garlic, spade, dandelion (pity the child whose birthday falls on dandelion day – the French word for dandelion is *pissenlit*, meaning 'bed-wetter'; also, the French have a saying, *sucer les pissenlits par la racine*, which means 'to suck the dandelions by the root', i.e. 'push up the daisies'), celery, endive, turkey, tobacco, beans, lentil, gooseberry, pitchfork, thyme, trout, and so on and so forth to the power of 360. Impossible to remember, this silly, affected, doffing gesture to the peasant classes was appreciated least of all by the people for whom it was designed. To them, *dimanche* (Sunday) was a holy day, the Lord's day; to supplant it with a load of donkeys and dandelions was heretical, so they continued to attend Mass on Sunday oblivious to the consequences. People caught dating documents in the old style were threatened with imprisonment.

To the 360 named days of the year were added five festival days that were called virtue, genius, opinion, labour (a particularly odd name for holiday!) and reward. This was a calendar which nobody could memorise, which inevitably tried the patience of the French citizenship, and which led to them eventually drifting back to the old, deeply ingrained names.

For primitive people, the day is the simplest and most obvious unit of time. Broken down into periods of darkness and light, sleep and wakefulness, it would be hard to imagine that even the earliest form of primate would not be able to recognise the day as a fixed, recurring duration, long before he had spotted the year, the lunar month, the recurrence of the four seasons, and long before he had learned to divide the day by numerical hours.

It is a curious fact that most languages of the world do not appear to have a word that differentiates between the day as a period of 24 hours (i.e. night plus day) and the day as measured by daylight hours. While we count our days from midnight to midnight, the Romans counted theirs from one dawn to the next; the Greeks did the same, as in the Homeric phrase: 'This is the twelfth dawn since I came to Ilium.' Caesar noted in his written account of the Gallic Wars: '*Spatia omnis temporis non numero dierum, sed noctium finiunt.*' Latin is a beautifully concise language, which is why it was used for so long as a cornerstone of education. English, though the most concise of the European languages, cannot match it. Is it possible to translate Caesar's nine words into any fewer than these 18 English ones: 'They define all spaces of time, not by the number of days, but by the number of nights'?

That Caesar should have been intrigued by this is surprising. Most people of the world count by nights. The Arabians talk about things occurring 'in three nights', as did their far distant ancestors who wrote on tablets in Sanskrit. For them, the period of 10 days is the *dacaratra* (*ratri* meaning 'night'). About the Britons, the Roman historian Tacitus (the most concise Latin writer of them all) managed to make the same point as Caesar had made about the Gauls, but added extra information and still managed to shorten the sentence length by a word!: '*Non dierum numerum, ut nos, sed noctium computant*' ('They do not count time by numbering days, as we do, but by counting in nights'). This custom has not been completely extinguished in Britain. The British still talk about a fortnight meaning a period of 14 days. The Taiwanese, Greenlanders, native American Indians and many other tribes and cultures counted nights and not days. The Dakota Indians counted days in 'sleep-times', the Kiowa measured the length of a journey in 'darks'. The reasons for this are not entirely clear, though it has been suggested that since the moon was such an important tool for measuring time, the night became one also – not an entirely convincing argument. Whatever the source, the practice of counting days from sunset to sunset is an ancient one. For Jews and Christians it is a tradition which has its roots in the Creation, a tradition which states, according to Genesis:

> And God saw the light and it was good: and God divided the light from the darkness. And God called the light Day and the darkness he called Night. And the evening and the morning were the first day.

In other versions, this has been translated so as to be even more explicit: 'And there was evening and there was morning, one day.' It is for this reason that the Jewish Sabbath (a tradition entirely based on the Book of Genesis) starts at sunset on Friday and lasts until nightfall on the following day.

In the modern dating system, known as standard time, the day begins at midnight all around the world, which means that any given day (say, 1 January 2007) will start at different moments in different places, depending where these places are on the earth's surface. Noon or midday (when the sun is at its highest point in the sky) should correspond to 12 o'clock wherever you are, which means that there must be more than one midday around the world during any period of 24 hours. The sun cannot be at its highest everywhere at the same time. So, if it is deemed to be 12 o'clock at whichever place the sun is at its highest in the sky, it stands to reason that, as the earth rotates on its axis, the time '12 o'clock' will move with it to correspond to whichever part of the globe is directly facing the sun.

Daniel Webster (1782–1852), that proud American, known to his childhood friends as 'Little Black Dan' because of his swarthy complexion, famously said in his speech on the completion of the Bunker Hill monument, 'Thank God, I – I also – am an American!' Webster's anti-British fervour was pumped to the brim at the thought that the British Empire once claimed territory in every time zone of the globe, as he made clear in his speech to the senate on 7 May 1834:

> On this question of principle, while actual suffering was yet afar off, the Colonies raised their flag against a power, to which, for purposes of foreign conquest and subjugation, Rome, in the height of her glory, is not to be compared; a power which has dotted over the surface of the whole globe with her possessions and military posts, whose morning-drumbeat, following the sun, and keeping company with the hours, circles the earth with one continuous and unbroken strain of the martial airs of England.

The earth, as we know, is carved up for man's convenience by lines of latitude (parallels) and longitude (meridians). Latitudinal lines run from east to west. The longest and most famous of these, the Equator, runs like a belt around the world's fat middle girth at an equal distance from both the South and the North Poles. The Equator is set at 0 degrees latitude. The South Pole is therefore 90 degrees south of the Equator and the North Pole is 90 degrees north. All other lines of latitude lie parallel to the Equator and consequently become shorter and shorter as their proximity to the poles increases. Latitude is used to identify a global position in terms of how far north or how far south it is.

For identifying a position to the east or the west we use the gridlines of longitude. These differ from the parallels in that they are all of the same length (i.e 12,500 miles, being half of the world's circumference), since all meridian lines pass through both the North and South Poles. The Prime Meridian is the line that passes through Greenwich. As the earth rotates, each meridian line will in turn face the sun, at which point it can be said to be midday on that meridian.

Before 1870 each country operated its own provincial system of local times. Large countries, like America, in which the sun is at its midday peak for five hours between its eastern and its western coasts, used to manage on a plethora of different local time zones, but the coming of the trans-American fast trains put an end to all that cowboy nonsense. Standard time was removed from the control of local town dignitaries and administered by an act passed by international consensus.

The present system was agreed at a conference of 27 nations which convened at Washington DC in 1884. Because there are 24 hours in the day, the world time zones were divided by 24 meridians, each separated by 15 degrees. The meridian line itself was deemed to be in the centre of each time zone. This allowed midday to pass, every hour, from one meridian time zone to the next, according to each 15-degree rotation of the earth. There was still a problem that needed to be solved. If midday is happening somewhere in the world at any given time, then so too is midnight (at the other side of the world – 180 degrees east or west), and since every new day of the calendar is supposed to start after midnight where, and at which midnight, is a new day supposed to start?

Before the International Date Line was set up, a traveller going right around the world, moving his clock either forwards or backwards as he entered each new time zone and moving his calendar forwards at 12 o'clock each night, would find, on his return, that the number of days he had experienced was different by one day from that which had been measured by the people left behind at the starting point.

To settle these irregularities, it was agreed that the new calendar day should dawn on the 180th meridian known as the International Date Line. The 180th meridian was chosen because it passes mainly through the Pacific Ocean. Drawing an International Date Line through the middle of a country would have been hopelessly inconvenient, for when it is midnight a day ends on the eastern side and a new day begins on the western side. This would have the effect of putting two neighbours in the same town on different calendar days from each other. As it happens, the International Date Line zigzags to avoid doing just that.

In the position where the International Date Line now is, the people who live on the Marshall Islands or in New Zealand (both just west of the International Date Line) will be the first to experience the new day. At 12.01 a.m. in the Marshall Islands on Friday 1 January 2027 it will still be 12.01 p.m. (just past midday) in Greenwich, London, on the previous day (i.e. Thursday 31 December 2026).

The International Date Line is a strange anomaly but at least it is a system that works. It even allows the superstitious or mad to miss out on days altogether without having to make a complete circle of the world. If you are scared stiff of what might happen to you on any Friday the thirteenth, here's a tip. Take a boat to the International Date Line and ask the captain to weigh anchor so that the deck of the boat is positioned along the 180th meridian. On Thursday the twelfth, just before midnight, stand, poised and ready on the east side of the meridian. You must time your westerly jump so that, at the exact point of midnight, you are flying over the meridian; when you land it will be Saturday the fourteenth. You can relax, you do not have to keep running and running, just ask the captain to sail off in a westerly direction and you can sunbathe on deck, sipping Coke, in the safe knowledge that Friday the thirteenth will never bother you again. It is as well to get yourself clear of the Date Line, though. If your boat capsizes, strong currents could blow you back over the line, where the dreaded Friday the thirteenth will still be lurking. In this event you are almost sure to be swallowed by a shark.

CHAPTER VI

SEPTIMANAE
Weeks

The subdivision of the months into weeks was one small but inevitable step for mankind. As a species we will insist on dividing and subdividing anything and everything until we are blue in the face. It is a peculiar human neurosis; no other species on earth, as far as we know, has shown the slightest interest in measuring or has ever expressed a wish to ally a number to any natural phenomenon. The human invention of numbers has, as we shall see, enabled mankind to hide himself away behind a screen of illusion. That is not to say that numbering systems have not had their uses. The invention (if one can call it that) of the week was not the upshot of a single piece of Roman imperial legislation but an inevitable process that grew from initial necessity.

Even so, the week is an artificial unit of time. It could have been five, six, seven, even 12 days long – what would it have mattered? Well, it would have mattered to some. We like to believe that our calendar is a smooth and competent machine, that within the space of one year we can fit an exact number of months, an exact number of weeks and an exact number of days into it. Unfortunately, that is not the case, though we would dearly love it to be so. Fitting lunar months into solar years requires fractional division. Days will not fit exactly into lunar months either, and nor, for that matter, can you get an exact whole number of days into a solar year – why should you? It would have been spectacularly strange if the amount of time it took for the moon to orbit the earth just happened to be an exact multiple of the duration of the earth's rotation on its own axis. Unsurprisingly, there are very few incidents we can point to in which nature is proven to be mathematically helpful to man. The hexagonal honey cells in a beehive are an exception, as are snowflakes and certain crystals, but as a rule the place to look for mathematical regularity is not in the natural world.

If God had been more considerate (or mankind less mad – it depends on which way you choose to look at it) he would have foreseen our human need for a simple, unfussy calendar and organised the revolutions of his solar system accordingly. Why were we not given a metric solar system, for instance? There could have been exactly 100 days to the solar year and exactly 10 days to the lunar month, making a neat 10 lunar months to the year. That would have sorted a whole lot of problems; and while he was doing that, he could have organised the earth to orbit the sun in a geometrically perfect circle without tilting, as it does, at the poles – this useful consideration would have enabled us to forget all about seasons, short days and long days and treat each day the same.

None of these things are as we would ideally wish them to be. The ancients searched for mathematical regularities, convinced that God was really interested in maths, but to no avail. Even in the early sixteenth century, Johannes Kepler, a brilliant scientist who later calculated the elliptical orbits of the planets, was trying to prove that the planetary orbits were based on combinations of geometrically perfect solids. All rubbish, of course. The planets themselves are not even properly spherical. God doesn't give a fig about maths, as we can all plainly see by looking out of the window into our gardens. We have been given a mathematically imperfect world, but the question we must now ask ourselves is whether we have made the best of the crazy system that has been allotted to us?

Yes or no?

We have inherited a year that is just over 365 days long. How are we going to divide it up into months and weeks or, to put the question back into the mouths of the ancients, 'How did God intend us to divide up this awkward batch of days?' A convenient week length is surely made up of a number of days that will divide exactly into 365. Five would have worked. We could have had a 73-week year made up of five-day weeks. But that would have been our only choice apart from seven, for no other practical number divides into 365. Early Teutons and some Hassidic races are known to have used the five-day week, but they were the exceptions. The seven-day week and 52-week year are as old and as prevalent as civilisation.

The great spur for the seven-day week seems to date back at least 3,000 years to the enthusiasm of the ancient Babylonians. It was from the Babylonians, as we have discovered, that the Jews, captives in their land, borrowed the story of Genesis. That God created the world in seven days gave the seven-day week a sacred significance to these peoples.

Thus the heavens and the earth were finished, and all the host of them. And on the seventh day God ended his work which he had made; and he rested on the seventh day from all his work which he had made. And God blessed the seventh day, and sanctified it: because that in it he had rested from all his work which God created and made.

And from this it is one easy step to Moses and the Ten Commandments from the Book of Deuteronomy:

Six days thou shalt labour, and do all thy work. But the seventh day is the Sabbath of the Lord thy God: in it thou shalt not do any work, thou, nor thy son, nor thy daughter, nor thy manservant, nor thy maidservant, nor thine ox, nor thine ass, nor any of thy cattle, nor thy stranger that is within thy gates; that thy manservant and thy maidservant may rest as well as thou. And remember that thou was a servant in the land of Egypt, and that the Lord thy God brought thee out thence through a mighty hand and by a stretched-out arm: therefore the Lord thy God commanded thee to keep the Sabbath day.

The sacredness of the Sabbath as a day of rest is something which orthodox Jews still take very seriously. It is, after all, a joyful reminder of their perpetual Covenant with God. But in the past it was taken so seriously that the Israelites would not even take up arms on a Sabbath – a useful tip for their enemies, who consequently slaughtered them without resistance. Rather than face extinction, new laws were introduced in the Talmud which allowed for the suspension of 39 forbidden categories of work when life or health was seriously endangered, for 'the Sabbath was given to man, not man to the Sabbath'.

So the main reason for the seven-day week seems to come from the Genesis story, but that still does not resolve the question as to why God chose to create the world in six days and rest himself on the seventh. If God is omnipotent, as we are so frequently reminded that he is, could he not have created the world in one day, or even in one second? But if the Creation really did need to take six days, why did he not spread it out a little; one day on, one day off; take a fortnight over it, conserve his energy instead of doing it all at once and then having to have a day of rest at the end? These might seem like strange, even banal, questions, but how else does one get to the bottom of things?

It is recorded that the ancient Babylonians were superstitious about certain days of the lunar cycle; cooking, travelling and marketeering were explicitly banned on these days. They were the seventh, fourteenth, nineteenth, twenty-first and twenty-eighth days of the lunar cycle. Pythagoras also feared the seventh day. Whether he took this from the Babylonians, or they got it from him, or the superstitions were independent, remains uncertain. What is interesting, though, is the connection between the sanctity of the seventh day in Jewish belief and the horror of it to the equally superstitious Babylonians. It suggests that the whole seventh-day resting idea of the Jews dates from their ignominious captivity in Babylonia, 600 BC. If this is so, it is a curious irony that unlucky days for the Babylonians should have been translated to lucky days by the Israelites. The Israelites, it will be remembered, fled from Egypt over the Sea of Reeds, a papyrus lake (not the Red Sea), under the leadership of Moshe (Moses), some time during the reign of the Egyptian Pharaoh Ramses II (1304–1237 BC). It has been suggested that Moses himself was the author of the Pentateuch (the collective name for the first five books of the Bible – Genesis, Exodus, Leviticus, Numbers and Deuteronomy). If we believe this (and many Hebrew scholars would dearly like us to) then the books are very old indeed. The oldest surviving written Old Testament is the Septuagint, a Greek translation from Hebrew made for Greek-speaking Jews in Alexandria between 200 and 100 BC. In 1947 a young goatherd stumbled on hundreds of Hebrew scrolls in 11 caves at Qumran on the Dead Sea. This astonishing find included fragments from every book of the Old Testament except one. They had lain undisturbed for nearly 2,000 years. The Dead Sea Scrolls included material which was older than any previously known biblical document, dating back to 300 BC. Unfortunately, all the fun and romance of this find has been destroyed by the fishy scholars who took the scrolls into their care and have never, even to this day, allowed the full texts to be shown to the world. It is thought that they may contain passages that cast doubt on the veracity of large sections of the Bible.

We have been allowed to see some of it, though. Most of these fragments would have been dated to some 200 years after the end of the Jewish captivity in Babylonia. We can surely dismiss, then, the theory that Moses was the author of the Pentateuch, since it contains too much that is clearly borrowed from the Babylonians and, as far as the Jews were concerned, the whole Babylonian episode took place at least 600 or 700 years after Moses' death. Is it possible, then, that the Babylonians, superstitious of the seventh, fourteenth, twenty-first and twenty-eighth days of the lunar cycle, ordered their slaves, the Jews, to do all the dirty work on those days for them? This might explain why, when they were finally freed from Babylonian captivity, the Jews set up their stringent rules about never again working on a Sabbath. They were so overjoyed to be freed from the ignominy of having to slave away for the Babylonians on these 'dirty days' that they invented a new set of rules: 'Thy manservant and thy maidservant may rest as well as thou.'

This is one possible interpretation, but we also know that in the earliest years of Jewish monotheism many of the laws of Moses were created solely in order to set the Israelites apart from other cultures, particularly those under which they had been held captive. The prohibition of graven images, for instance, has been interpreted as a deliberate reaction against Egyptian belief. Similarly, the laws which prevented the Israelites from eating young goats stewed in nanny goats' milk (Exodus xxii, 19) or from trimming their hair at the edges (Leviticus xix, 27) were passed to distinguish the Israelites from their mortal enemies, the Canaanites, and to insult the latter by prohibiting what they held to be either sacred or desirable. In this light, the Jewish reverence for the number seven can be interpreted as a planned reaction against the abhorrence of seven on the part of the Egyptians and Babylonians. As the Roman historian Tacitus pointed out of the Jews: '*profana illic omnia quae apud nos sacra, rursum concessa apud illos quae nobis incesta.*' (They consider all that is sacred to us profane and promote everything that for us is prohibited.)

But why did the Babylonians (as well as the Egyptians and the Sumerians) decide that seven was such an evil number in the first place? Simple: these ancient peoples, as we have already seen, endowed the number 60 with a sacred significance. Smaller numbers that divide into 60 were consequently held to partake (to some extent) in the holiness of this number. Thus, counting up from zero, the numbers 1, 2, 3, 4, 5 and 6, which all divide into 60, were considered goodies while seven – the first of the sequence which will not divide into 60 – was reviled as a dirty, rotten bringer of ill-fortune.

In any event Saturday, which we now know as the seventh day, is named after the god Saturn, and both Saturn the god and Saturn the planet were long considered the bringers of ill-fortune. If that link existed in ancient Babylon, and there is no reason why it might not have emanated from there in the first place, then we have our explanation as to why the superstitious Babylonians were reluctant to do anything that required good fortune on the seventh day. For whatever reason the Jewish seven-day week was well entrenched at the height of the Roman Empire. It is mentioned as a point of interest by Ovid, Seneca, Juvenal and others when writing about the Jews.

Whether the Jews invented the week or whether it was taken by them from the Babylonians remains frustratingly unclear, but what we do know is that the Jewish week began on the day after the Sabbath. In AD 321 Emperor Constantine, who embraced the Christian religion, declared the day after the Sabbath to be the Lord's day (*Dies Domenicus* – 'Sunday') and that day was similarly declared the first day of the Roman week. For the Jews the number seven became something of an obsession. They had three sorts of 'weeks': the first, the common one of seven days; the second was of seven years; and the third one was of seven times seven years, at the end of which was the jubilee. The Jewish jubilee was celebrated every fiftieth year to commemorate their safe deliverance from Egypt. The name is now used to indicate almost any fiftieth anniversary event. 1887 was the year of Queen Victoria's Jubilee, which afforded the public an opportunity to buy mugs and scarves for which they had no need. So many were made that their value has decreased since the year of their manufacture. Elizabeth II's Silver Jubilee was celebrated in the twenty-fifth year of her reign, and Victoria's Diamond Jubilee of 1897 marked 60 long years on the throne. The word comes from the Hebrew *jobil* which means 'ram's horn':

And thou shalt number seven sabbaths of years, unto thee, seven times seven years; and the space of the seven sabbaths of years shall be unto thee forty and nine years. Then shalt thou cause the trumpet of the jubile to sound on the tenth day of the seventh month, in the day of atonement shall ye make the trumpet sound throughout all your land. And ye shall hallow the fiftieth year, and proclaim liberty unto all the land and unto all the inhabitants thereof: it shall be a jubile unto you; and ye shall return every man unto his possession, and ye shall return every man unto his family. A jubile shall that fiftieth year be unto you: ye shall not sow, neither reap that which groweth of itself in it, nor gather the grapes in it of thine vine undressed. For it is the jubile; it shall be holy unto you: ye shall eat the increase therof, out of the field.

Leviticus xxv, 8–12

David Hume (1711–76), the pleasantest and most urbane of all philosophers, would have had an answer for that:

If we take in our hand any volume; of divinity or school metaphysics, for instance; let us ask, *Does it contain any abstract reasoning concerning quantity or number?* No. *Does it contain any experimental reasoning, concerning matter of fact and existence?* No. Commit it then to the flames: for it can contain nothing but sophistry and illusion.

The origins of the word 'week' are somewhat obscure. Even the best etymological dictionaries struggle with it, suggesting only that it may have had its origins in words meaning 'succession', 'series', 'bend', 'turn', 'move' or 'change'. Interestingly, the word 'week', which is not related to the French, Italian or Spanish (respectively *semaine, settimana* and *semana*), is common to the Scandinavian, Teutonic and Yiddish languages. It is *woche* in German, *vecka* in Swedish and *vokh* in Yiddish. Might the Israelites have snitched that from the Babylonians as well? Who knows?

The Romans carved up their months in two different ways. There was the seven-day week, similar to the modern African market week (see page 84), which named its days after the seven planets, and then there was a loftier system based on *kalends*, *nones* and *ides* drawn from the phases of the moon. This was the system used by the holy pontifices and consequently it became the official means by which all dates were recorded on documents, announcements and memorials.

The first day of the month, in other words the new moon, was called the *kalends* (whence our word 'calendar'), which one Roman scholar, Marcus Terentius Varro (116–27 BC), says originated in the practice of 'calling' the people to the temple for sacrifices at the new moon. The *nones* was the ninth day before the *ides*, and the *ides* was originally supposed to be the day of the full moon, but because the Romans had indulged a superstitious terror of even numbers the *nones* was ascribed to the fifth or the seventh of each month and the *ides* to the fifteenth or the seventeenth. This system of weeks was hopelessly impractical since nobody knew on what days the festivals would fall – nor *could* they know, unless they were within earshot of the temple from where a junior pontiff was supposed to yell '*calo*' at the top of his voice. The idea was that he would call several times, each shout of '*calo*' representing one day and indicating, to anybody who was listening and/or bothering to count, how many days would elapse before the opening day of the next month. The self-important pontifices, seeing it all as part of their sacred obligation, tried to hold on to the tradition for as long as they could. Unfortunately, few Roman citizens took the blindest bit of notice, generally operating on a market-week system inherited from the Etruscans. The calling of the *kalends* was eventually dropped in 303 BC and the *fasti*, or calendars, were thereafter posted up in public places.

The *ides* are famous because of the warning prophesy to Julius Caesar: 'Beware the *ides* of March.' Being an intensely superstitious Roman, Caesar paid great attention to his soothsayer's warning and kept himself from trouble on the *ides* of March, but since nobody, except the pontifices, understood the system properly, Caesar appears to have been a trifle muddled as to when the *ides* were over, as Plutarch's life of the emperor confirms:

Furthermore there was a certain soothsayer that had given

Caesar warning long time afore, to take heed of the day of the *ides* of March (which is the fifteenth of the month), for on that day he should be in great danger. That day being come, Caesar, going into the senate house and speaking merrily unto the soothsayer, told him 'The *ides* of March be come'; 'So be they,' softly answered the soothsayer, 'but yet are they not passed.'

Caesar promptly gulped a mouthful of spittle and looked towards his sundial. 'Eheu! The soothsayer had a point!' Within the hour Caesar was dead.

The Roman numbering system works in a cumbersome fashion, involving a laborious operation of counting both forwards and backwards. Take a simple number, like the date 1998, for instance, and it becomes instantly obvious that the Roman system is a dinosaur because 1998 looks like this: MCMXCVIII (where M is 1,000; C is 100; X is 10; V is five and I is one). The number itself (1998) is suggested by the following compound sum: M (1,000) plus CM (1000 minus 100, i.e. 900) plus XC (100 minus 10, i.e. 90) plus V (five) plus III (one plus one plus one, i.e. three) = 1998. The same hopeless procedure of backward counting was introduced into the official calendar; for instance, 6 January would be called 'a.d. VIII. Id. Jan', which literally translated as a.d. (*ante diem* – day before) VIII (five plus one plus one plus one, i.e. eight), Id. (*ides*) Jan (*Januaris*), in other words eight days before the *ides* of January. In the same way 3 March would be 'a.d. III. Non. Mar', and 29 January would be 'a.d. IV. Kal. Feb' (i.e. the fourth day before the *kalends* of February).

The Chinese calendar has weeks of both 10 and 12 days in duration. In the 12-day weeks the days are named after animals – the 12 terrestrial branches, *chih (zhi)* – while in the 10-day week they are named after the 10 heavenly stems *kan (gan)*. This alternating system, known as the *kan-chih* day-count system, has been in use for more than 3,000 years.

The Egyptians used a system of 10-day weeks called *decans*, which were copied by the French with the impractical system that was enforced after the Revolution of 1789. Some of the oldest weeks are based not on the stars but on market sharing systems between local villages. This is particularly the case in Africa, where a number of villages will share a market which is set up in a different village each day following a strict rota. The length of the week depends, in these instances, on the number of participating villages.

The Bakongo tribe of Central West Africa has four market villages, called *konzo*, *nkenge*, *usona* and *nkandu*, and these names are in turn given to the four days that comprise the Congo week. This scheme is common right across Africa and on certain islands in the East Indies. The Mexicans used to have 13-day weeks, named after the 13 lords of the heavens, alternating with a nine-day week, named after the nine lords of the night. Most complicated of all is the Javanese week, which has been in use among the remote tribes of Indonesia since the tenth century. Their system, which consists of nine different cycles (a two-day cycle, a three-day cycle, a four-day cycle and so forth, up to 10), takes its origins from sacred rites and the ancient belief that each number is pregnant with a mystical religious significance.

The ancient Greeks went along with the seven-day week for reasons unconnected with the Jewish penchant for the number seven. It was entirely a planetary consideration for the Greeks. They knew of the seven planets and, like the Romans, ascribed gods to each of them, so that each god ended up with a planet and a day of the week to patronise. Kronos, god of time to the pre-Hellenic population of Greece, was set by these people to monitor the whole week. But Kronos was a revolting figure, a boastful Grobian oaf, similar in his most unappealing attributes to the Roman god Saturn. On Greek vases and psykters, Kronos is depicted as an old man with wide, beady eyes and carrying a sickle, or in later images carrying a curved sword called a *harpe*. That later Greeks reviled him is borne out by the inhuman grimace with which he is depicted. The harvest festival of Kronia, which mirrored the Roman Saturnalia, was a chance for all Greeks, including slaves, to say and to do exactly as they pleased. Moral restrictions were lifted, and the festival was generally anticipated as the wildest event of the year, reflecting, no doubt, the extraordinarily bad behaviour of the guardian of their week, Kronos.

His deplorable actions started when, on the advice of his mother, Kronos castrated his father with his *harpe*. Uranus was his father, Gaea his mother. They respectively represented heaven and earth. Kronos's paternal castration was thus a symbol of the separation of heaven from earth in the Greek version of Genesis. Like the Sumerian and Babylonian gods, Kronos was also invested with a predilection for incest. By his sister Rhea he had five children, whom he named Hestia, Demeter, Hera, Hades and Poseidon. Like a modern crocodile, he swallowed the lot. His next child, Zeus, was jealously protected by his mother/aunt and kept hidden from the greedy Kronos. Desperate to put the young Zeus down his gullet, Kronos lurched from pillar to post like a drunken buffalo with a rumbling tummy. At one moment he thought he saw a figure that he believed to be his son and swallowed it. But Rhea had tricked him into swallowing a stone. Later he sicked up the five children he had swallowed (all of them miraculously shot out alive) and was defeated by Zeus in a pitched battle at Mount Olympus. The Greeks are muddled as to what happened next. In some versions, Kronos is held prisoner in Tartarus; in others he is made King of the Golden Age.

Why this terrible ogre was allowed to govern the Greek harvest as well as the Greek week defies rational explanation. On the whole the Greeks tolerated bad gods for they offered an explanation for all the evil in the world. In the Christian and other monotheistic religions the ideal of a good god is severely put to the test by the presence of so much evil. 'If God is so good and he made everything, why is there evil?' This question continues to fox clergymen to this day. The ancient Greeks had the luxury of Kronos to blame for all their woes. We comfort ourselves with the thought that he was a reviled figure and the least worshipped god of all.

CHAPTER VII

MENSES
Months

The moon is the most intriguing planet in our celestial plain; we spin our imaginations to define it – 'the horned moon', 'that orbed maiden, with white face laden', 'that cold and fruitless moon ... the governess of floods' or, as Christopher Fry put it in his typically polysyllabic way: 'The moon is nothing but a circumambulating aphrodisiac, divinely subsidised to provoke the world into a rising birth rate.' For others it has been seen as a balm, a lantern, an old man, a goblin, a wild goat, a piece of cheese, a lady's bottom and, by e. e. cummings, who pondered in an optimistic vein: 'who knows if the moon's a balloon, coming out of a keen city in the sky – filled with pretty people?'

The sun, by contrast, has never inspired such a wide range of descriptive epithet. For starters, it is far too bright and it hurts us to look at it. We cannot lie on our backs gazing up as we do with the moon. In past times the sun was worshipped as Helios, protected by the great god Apollo; to the Egyptians the sun was Ra; Shamash to Assyrians; Meradoch to the Chaldeans; Ormuzd to the Persians; Tezcatlipoca to the Mexicans; and, more straightforwardly, Sunna to Scandinavians. Not many of us bother to worship the sun any more (though there are believed to be little enclaves of modern, sandal-wearing sun-disciples in America). Most of us have forgotten what sex he or she is traditionally supposed to be. Which is not to suggest that the human race has become any more rational or admirable since those ancient days. If our distant ancestors could have looked forward in time and seen our modern ways, they would have doubtless concluded that we, with our politically correct attitudes, are all completely crazy. To an ancient observer, modern man's relationship with the sun might be defined thus:

Its harmful rays, he is told, will cause his offspring to suffer skin cancer. The school-teacher who forgets to provide sun-block to his pupils will be sued; the teacher who applies the cream to their skins will be arrested, and his computer will be confiscated; the child who applies the sun-block to his own skin will be taken to a counsellor; the child who spills it on the ground will kill an insect, and the manufacturer will be held accountable either for the death of the insect or for some other unforeseen calamity which his unction is deemed to have caused.

Mars is not easy to spot with the naked eye. Though less demonstrative than the sun, it has long been a planet of hostile connotations. It was called Mars because of its reddish colour, and red is the colour of blood, blood is the symbol of battle and Mars is the Roman god of war. Mars is an untrustworthy planet. Underneath it, according to the *Compost of Ptholomeus*,

> is borne theves and robbers, nyght walkers and quarell pykers, bosters, mockers, and skoffers; these men of Mars causeth warre, and murther and batayle. They wyll be gladly smythes or workers of yron, lyers, gret swerers. He is red and angry, a maker of swordes and knyves, and a sheder of mannes blode . . . and good to be a barboure and a blode letter, and to drawe tethe.

Mercury is even harder to spot because of its proximity to the sun. The fleet-footed planet flies through the night sky, taking only 88 days to make a complete solar orbit. Mercury, like Mars, casts a sinister shadow. It was always thought that crooks were born under it, as Shakespeare observes: 'My father named me

Autolycus; who being, as I am, littered under Mercury, was likewise a snapper up of unconsidered trifles.' Venus (which according to Chaucer 'loveth ryot and dispense'), Saturn and Jupiter (both symbolic of chaos) were all the planets known to the ancients. Each can be seen without the aid of a telescope, but you need an expert to point them out. The moon offers the non-expert, the amateur stargazer, a chance dream . . . It has a beaming face – only a fool could miss it – and only a fool would fail to notice that it is obviously spherical (as opposed to flat and disc-like, which in ancient times all the other planets, including Earth, were deemed to be). The moon can be scrutinised in great detail with the naked eye and, while its sudden appearances in the middle of the night sky have, on occasion, taken the unlearned observer by surprise, its pattern of waxing, its waning and its regular 29½-day cycle of phases have been comforting and predictable to the dimmest teenaged shepherd since the dawn of time. While most ancient astronomical wisdom was controlled by the kings and their high priests, the moon gave everybody a chance to share in the power; to fantasise and mythologise and to lie in its light, pondering the meaning of life. Hardly surprising, then, that so many fables and legends have sprung up about the moon. There has been more talk about it than about any of the other celestial bodies, including the sun.

By tradition the moon is a masculine satellite planet and the sun is female. The French have it the other way around, but in the Icelandic *Edda* books (two collections of twelfth-century mythological poems discovered by a bishop in 1643 and errone-ously attributed to Saemund Sigfusson) the moon is Mani, son of Mundilfoeri and his sister is Sol, the sun. Lithuanians and Arabs still sex the sun and moon in this way, while in German folklore the two 'great lights' are jovially referred to as Frau Sonne (Mrs Sun) and Herr Mond (Mr Moon). The German Mr Moon is not to be confused with that other figure of popular folklore, 'the man in the moon'. To the Teutonic mind Herr Mond represents the moonification of Teutonic man and is emblematic of all that is best in the German male. Herr Mond is a sturdy, jovial fellow, disposed to philosophical contemplation; sentimental towards children; a man who enjoys the act of laughing more than the joke that was supposed to have provoked it. He brings presents at Easter, and on German television his beaming beacon of a face is used to promote cloudy wheat beer, smoked cheese and tasty, frozen schnitzels.

The man in the moon is an altogether more enigmatic character than Herr Mond, who would be quite useless to the advertising industry. Looking at the moon we can see blotches and marks, deep craters too, some of which were formed by crashing meteors millions of years ago. The unfocused or the inebriated eye will make out patterns between these dark and light patches, seeing a nose here, a soup-dish mouth there. Some observers claim to see a Highland terrier with a bone in its jaws and a patch over one eye. Over two images, though, there appears to be some consensus. Both of them are known as the man in the moon for the reason that they both look like men, but it is only ever possible to see one of them at a time. The most obvious is a face, a grumpy face in profile, not without intelligence, with a sallow cheek and deep-set, judgemental eyes. The other takes the form of a man who seems to be leaning against something, possibly a pitchfork. One idea is that he is the man from the Old Testament who had the misfortune to be caught gathering sticks in the wilderness on a Sabbath. Strictly speaking, picking up sticks counts as work, and work was and still is forbidden by Jews on a Sabbath. In this particular story, which appears in the Book of Numbers, some goody-goody zealots, describing themselves as 'Children of Israel', take it upon themselves to report the man's stick gathering to Moses:

And the Lord said unto Moses; The man shall be surely put to death: all the congregation shall stone him with stones without the camp. And all the congregation brought him without the camp, and stoned him with stones, and he died; as the Lord commanded Moses.

As Joseph Conrad put it: 'There is nothing so dangerous as a fanatic convinced by the justice of his cause.' It would be comforting to think that if the man in the moon were indeed this same unfortunate stick-gatherer, he shall for ever continue to stare down at this earth to serve as a haunting reminder of the mob rule and bloodshed which results from uncontrolled religious zeal. Neither God, nor Moses, nor the children of Israel emerge from this tale with any credit to themselves. Let us not forget that embarrassing passage from the Book of the Prophet Isaiah: 'I form the light and create darkness: I make peace and create evil: I the Lord do all these things.' In the Jewish creed the word 'evil' is changed to 'all things'.

Be that as it may, in the prologue to *A Midsummer Night's Dream*, Shakespeare tells us that 'this man with lantern, dog and bush of thorn, presenteth moonshine'. In other accounts the man is Cain, the fallen son of Adam, while to many poets he is Endymion, the beautiful youth who slept on a ridge at Mount Latmus, exciting the passions of the cold moon-goddess Selene so much so that she came down to earth and kissed him. This version of events forms the basis of John Lyly's sixteenth-century comedy, *Endimion, the Man in the Moone*.

And to confuse the issue further, here is an old English rhyme:

> The man in the moon
> Came down too soon
> And made his way to Norwich.
> He went by the south
> And burned his mouth
> By eating cold plum porridge.

Many languages, including our own, derive their word for 'moon' from the ancient linguistic stem 'me'. The Aryan is *moanh*, the Sanskrit *mas*; it is *men* in Greek, *mi* in Irish, *mond* in German, *maan* in Dutch, *maane* in both Swedish and Danish, and *metzle* to the Mexicans. The list could continue, but the point is made. From the 'me' stem comes the English verb 'to measure', as does the Latin *mensis* (a month) and, of course, the word 'month' itself.

The Latin name for the moon, *luna*, is colloquially *lucna*, which in turn is derived from *lucere*, to shine, and is consequently the basis for our English word 'lunatic'. The Romans started off the rumour that people's behaviour becomes stranger and stranger towards the middle of each month, culminating with wildness and folly at full moon. This rumour has persisted since the days of Caesar, despite many efforts through the centuries to dispel it. Listed as one of Crozier's Popular Errors, and long believed to be fictitious, the connection between mental instability and the waxing moon was given a new lease of life by an English social survey conducted by the Medical Association in 1999. The survey showed that the worst incidents of prison rioting always occurred at full moon.

While the moon fuelled so many wide-ranging mythologies, it appears to have been used from the very earliest years of human civilisation as a measuring tool for time. The first calendars, or what we believe might have been calendars, date from between 13,000 and 30,000 years ago. They are short animal bones with notches chiselled out of them in regular intervals and marked into groups. One of them, a Cro-Magnon eagle bone, discovered at Le Placard in southern France, provides evidence to suggest that the moon was the earliest celestial tool used by man for timekeeping.

Cro-Magnon man was first discovered in 1868 buried in a shallow cave on a hill called Cro-Magnon in the Dordogne area of France. He and his race were of a stocky build, growing up to five feet seven inches (1.7 metres), and their brains were bigger than ours. They have been honoured as the inventors of art. Gifted pictures of large-breasted, pregnant women and a range of wild and domesticated animals adorned the walls of their cave dwellings, both in southern France and in Spain. These paintings should not be compared in the skill of their execution with the great works of the last 2,000 years, but art critics have nevertheless freely lauded them as 'surpassingly beautiful'. The Cro-Magnon's sensitive aesthetic spirit led him to decorate his tools and weapons, bury his dead and to live in fixed abodes. By 10,000 BP (Before the Present) he seems to have merged with other European races: either that or his line went dead, we cannot be sure.

Cro-Magnon man was interested in the phases of the moon

as observed from his vantage point in southern France. For Cro-Magnon man, the phases must have been easy to follow in a cloudless month. He would have noticed the new, the crescent, the half, the gibbous (more than half) and the full, recurring in a regular sequence each and every month. He might not have been aware that the whole process, from one new moon to the next, takes precisely 29.530588 mean solar days, or 29 days, 12 hours, 44 minutes and three seconds. It was a rougher version of this time-span that allowed him to measure the earliest units of calendrical time. But Cro-Magnon still had much to learn. It was many thousands of years later that the Greek astronomer Meton observed that the new moon falls on the same day each year according to a 19-year cycle. A hundred years later, in 330 BC, Calippus, another learned astronomer, realised that the Metonic cycle was not quite perfect – it worked only if you dropped a day once every 76 years – but all of this would have gone right over the head of Cro-Magnon.

What we now know is that the moon orbits the earth, not in a perfect circle but following the trajectory of an ellipse while its rotation is tilted at a five-degree angle to the sun. These differentials allow us to measure the month in many different ways. Most of them need not concern us here; they are fodder for astronomers and wearisome to explain. The 29½-day month is the one that Cro-Magnon man would have observed at Le Placard and is about two days longer than any of the other astronomical months. The reason for this is simple. What Cro-Magnon man was counting was the synodic month, the time it takes for the moon to orbit the earth and return to the same position relative to the sun. Because the earth is making its own gigantic loop of the sun at the same time as the moon is making its much smaller orbit of the earth, the moon has to travel more than 360 degrees around the earth before it reaches the same point of alignment with the sun. If this were not the case, and the earth were perfectly still, the moon would be able to orbit the earth over 360 degrees, taking only 27 days and eight hours. These 29½-day cycles have formed the basis for all calendrical months.

It was only a matter of time before the months were given their own names. Back to the Sumerians, our ancestors from the font of all civilisation. They had not managed, or perhaps never tried, to calculate the length of the year by observing the ways of the sun, yet it was obvious to them that after roughly 12 lunar cycles the seasons started repeating themselves. On this basis they were able to give names to all of the months. Of course, without the intricate system of months of different lengths and leap years that we presently have, the months gradually slipped through the seasons. What started as *se-kin-kud* (month of the corn harvest), *apin-du-a* (month of opening the irrigation pipes) and *su-kul-na* (sowing month) must have been grossly misnamed after a period of only 15 years. Many civilisations (the Sumerians were the first) adopted calendars based on lunar phases, only to find that the names they gave to the months needed constant updating as they slowly drifted against the seasons.

The names we presently use have remained more or less fixed for the last 2,000 years, and the reason for this is that instead of being named after the seasons (Corn Month, Snow Month, etc.) they have taken the names of Roman gods, emperors and numbers, allowing the months to slip against the seasons without it ever seeming to matter. Ironically, our ninth, tenth, eleventh and twelfth months (September, October, November and December) are named after Roman months meaning 'seventh', 'eighth', 'ninth' and 'tenth'. But since so few people are able to speak Latin nowadays, it hardly seems to matter.

The Western calendar that we use today has borrowed all of the names for its months from the Roman Empire and is, in this respect, a calendar of purely pagan origin. January, which was called after the Roman god Janus, alternated with March as the first month of each new year. Janus, a god with two faces, one on each side of his head, was the symbolic deity of the gateway. With his double set of eyes he was able to look simultaneously backwards to the old year and forwards to the new – and so it was that the Romans honoured him with the first month of the year. In times of war the Romans flung open the doors to the

Temple of Janus hoping that some of his double-sighted magic would waft over to the battlefield and help the troops to notice impending dangers on all sides at once.

In the old system that Julius Caesar inherited when he came to power in Rome during the first century BC, February was the last month of the year (March being the first) and also the shortest, with only 28 days. The name is taken from the Roman festival of purification, Februarius, which was held on the fifteenth of February each year, and, like most Roman festivals, it afforded an opportunity for the citizens to run amok, drinking and fighting in the streets, making loud showing-off recitations on the steps of the temple. The purification seemed to consist not of abstinence and confession in the Christian sense but a crude, orgiastic frenzy of animal sacrifice and offal-burning. Later, when February became the second month of the year it was straightaway considered a dangerous and doom-laden month. The Romans quaked in their tunics whenever the number two was mentioned. More on this later. The Dutch call February *Spokkelmaand*, which means 'vegetation month', remarkably similar to the old Saxon *Sprote-cal* from the sprouting of pot-wort or kale.

March is named after Mars, the god of war, which was in consequence a deeply mistrusted planet, the 'boster and quarell pyker' of the *Compost of Ptholomeus*. Evidence of a face on Mars has emerged after scrutiny of the planet's mountainous surface with modern satellite cameras. While the more po-faced astronomers have dismissed it as a geological fluke, others have promoted the image (which looks a bit like a thoughtful cat) as evidence of a lost Martian civilisation. More fanciful still is the American cult religion called Roots of Mars, which teaches its followers that human civilisation started on the red planet and that conditions out there became so cold and unpalatable that they had to make an escape – and escape they did, by crash-landing on earth, somewhere between the Euphrates River in the southern Sumer, 40,000 years ago. According to Roots of Mars we are their living descendants. Perhaps this explains why we have never found the missing link!

The Latin for 'to open' is *aperire*, and from this the month of April takes its name, referring to the opening of buds, the reappearance of green leaves and the general sense that the womb of nature is opening itself to the creation of new life. The same idea is reflected in the French Republican calendar which gave for April the name *Germinal* – the time for budding.

The myth of the April fool is a curious one, as it exists also in France as *un poisson d'avril*, in Scotland as a 'gowk', as well as in India as the *dahti*, exposed at the time of the holy Huli festival on 31 March. Efforts have been made to discover the origins of what, in all cases, involves an initial 'joke' of deceiving someone then propounding the fun of it by calling the deceived by a silly name. Two theories (that it is a symbolic re-enactment of the 'deceptive' April weather, and that it is a representation of Christ's trial by Pontius Pilate) have been dismissed. It is more likely that the April fool is derived from the Roman festival of Cerealia, which occurred every year at the beginning of April. Ceres, after whom the festival (and latterly breakfast cereal) was named, was the Roman equivalent of Demeter, Mother Earth and great protectress of the harvest. It was to Ceres that Roman farmers made their sacrifices and to Demeter that the Greeks turned for their agricultural assistance. When the infernal Pluto snatched Demeter's daughter, Proserpine, from a daffodil field, Demeter heard the screams and rushed to her daughter's aid, little realising that she was chasing an echo in the wrong direction. By the rules of the game, Pluto, at that point, should have cried 'April fool!' (or 'gowk!' if he had any Scottish blood in him) and returned Proserpine to her mother, and everyone would have had a good laugh about it. That is not what happened.

Demeter was overcome with grief when she learned that Pluto was refusing to release her daughter from the Underworld and, in her distress, forgot all about the harvest. When the crops failed and famine ensued, Zeus intervened, ordering Pluto to release Proserpine at once. However, it was decided that since Proserpine had enjoyed a single pomegranate seed in the Underworld, she should be bound to spend a third of her time in Hades and two thirds with her mother on Mount Olympus.

May is the month when many peoples of the earth (most particularly the English) make the greatest fools of themselves. The Romans were habitually making fools of themselves (despite their oft-repeated adage *Semel in anno licet insanere* – 'We may play the fool but once a year'), and so it will come as no surprise to learn that May, like all the other Roman months, provided these hedonistic ancients with yet another excuse for singing and dancing, drinking, eating, and vomiting all over their nice clean togas: all with the tacit approval of another god. This time it was Maia (from whom the word 'major' is derived), a goddess who, like Ceres, had the interests of farmers at heart, with her special area being quantity and profit. The same idea is obliquely reflected in the French Revolutionary month name *Floreal* and the Dutch *Blou-maand* ('blossoming month'). Roman youths took to the fields, skipping and dancing in honour of Maia and Flora (the goddess of flowers) at this time of year. In one of Europe's quaintest traditions, the English stick a pole into the ground with coloured ribbons knotted to the top. Each taking hold of a coloured string, they dance round and round the pole, occasionally changing direction or alternating hops with skips. The ancient ritual connotations with Maia and rich harvest have all but disappeared, though maypole-dancing is still continued in England by stockbrokers wishing to draw attention to themselves on their days off. May Day in England is also known as Robin Hood Day, because the famous outlaw of Sherwood Forest was supposed to have died on that day. The sight of a few men in green nylon tights is as much as you will see of it today.

An old myth from Poland ascribes a precious stone as guardian to each calendar month. The May stone is an emerald, representing success in love: an ironic choice given that another, even older May-time celebration was the Roman festival of Bona Dea, goddess of chastity (no doubt celebrated with wild orgies). This tradition made some Europeans windy about May marriages.

The Romans, for no known reason, also celebrated Lemuralia (the feast of the dead) in May.

To the Romans, May marriages were ill-advised, while marriages in June were positively welcomed. Juno provided Romulus, Rome's founding emperor, with a name for the month, and since she, Juno, the 'venerable ox-eyed' wife of Jupiter, was the official guardian of women and marriages, the Romans, who were otherwise short of festivals at this time of year, all chose to marry in June. Singing, dancing and wild frolicking were much to be encouraged by the leaders in Rome, for debauchery, it is said, always ensures a degree of political stability. 'The way to keep the mob in check is with bread and circuses,' wrote the Roman satirist Juvenal.

In most of Europe and America, June is still the favoured month for marriage, though one suspects that this has more to do with the weather and a growing desire for marquees, cold farmed salmon and champagne than with anything else.

July takes its name from the Emperor Julius Caesar, who, as we shall see, was one of the great calendar reformers of history. It was Caesar's idea to call the month after himself, but since he was regarded as something of a god, the vanity of his gesture should not be despised. The new name took effect in 44 BC, the year of his death. Many years later, the next month after July, originally called *Sextilis* as it was the sixth month from March (one-time first month of the year) changed its name to Augustus in honour of Caesar's successor as emperor, Caesar Augustus. Although Augustus's birthday was in September, *Sextilis* was the emperor's lucky month, as the resolution passed by senate at the time makes clear:

Whereas the Emperor Augustus Caesar, in the month of *Sextilis*, was first admitted to the consulate, and thrice entered the city in triumph, and in the same month the hostile legions from the *Janiculum* placed themselves under his auspices, and in the same month Egypt was brought under the authority of the Roman people, and in the same month an end was put to the civil wars; and whereas for these reasons the said month is, and has been, most fortunate to this empire, it is hereby decreed by the senate that the said month shall be called Augustus.

There was a problem, though. *Sextilis* had only 30 days in it, yet July had a fulsome 31. The implications were too horrible to contemplate. If Julius Caesar's month had 31 days and Caesar Augustus's only 30, history might conclude that Caesar had been the greater emperor. Keen to set the record straight, Augustus removed a day from February and transported it to his own new month of August. It is for this reason that February, to this day, remains the shortest month. To many historians Augustus was just as fine an emperor as Julius Caesar – equal in efficiency to his predecessor, a noted genius for administration, less proud, marginally better controlled in his sexual behaviour, more devoted to the citizens of Rome and, in consequence, marginally more popular with his people. As a ruler he has been widely praised, but he suffered from one incurable weakness to which many commentators have alluded – vanity. The naming of August by Augustus after himself was not an isolated example of this type of behaviour. He loved to look at himself in the mirror, excited by his own good looks and intrigued by the effect he had on others. He was also known to enjoy being painted and sculpted and adorning his rooms with figurative models of himself. But, for all this, Augustus was a man of the people; he eschewed luxury in favour of the quiet contemplation of nature. He loved flowers and birds, spending hours on end pruning and clipping in his fountained garden in Rome. A contemporary bronze sculpture from Meroe in the Sudan (now in the British Museum) shows him to have been justified in his own assessment of his looks – this and other statues were later used as models by the French sculptors of the much less attractive Napoleon Bonaparte. A look of astonishment across a wide-eyed face suggests that Augustus was an enthusiast of some sort, but Gaius Suetonius, his biographer, puts forward the idea that the great emperor was convinced that his eyes flashed a divine power:

> He was unusually handsome and exceedingly graceful at all periods of his life, though he cared nothing for personal adornment. His expression, whether in conversation or when he was silent, was calm and mild ... He had clear, bright eyes, in which he liked to have it thought that there was a kind of divine power, and it greatly pleased him, whenever he looked keenly at anyone, if he let his face fall as if before the radiance of the sun. His teeth were wide apart, small and ill-kept; his hair was slightly curly and inclining to golden; his eyebrows met ... His complexion was between dark and fair. He was short of stature, but this was concealed by the fine proportion and symmetry of his figure, and was noticeable only by comparison with some taller person standing beside him.

The remainder of the calendar months (September, October, November, December) were simply numbered and not named. They stand for seventh, eighth, ninth and tenth, referring to the old system of 10-month years which started in March. With Caesar first, and then Augustus, having months of the year named after them, it would not have taken a prophet to predict what would happen next. Of course, all the emperors wanted months with which to glorify their names. That matricidal aesthete Nero took *Aprilis* and changed it to *Neronius* to commemorate his safe escape from an assassination plot in AD 65; the club-footed, tongue-twisted Claudius had May; and Germanicus, adopted son of Tiberius, made off with June. Interestingly, Tiberius himself rejected a proposal from the senate to change the name of September to *Tiberius* and October to *Livius*. 'What will you do when there are 13 Caesars?' he is quoted as asking. Outside Rome, provincial leaders eager to flatter their emperor piled honorific names into the calendar as a matter of course. At one time the Cypriot calendar had months named after Emperor

Augustus; his son-in-law Agrippa; Livia, his wife; Octavia, his half-sister; and Nero and Drusus, his stepsons. The same calendar also included a month in honour of the mythological Trojan hero Aeneas who, according to Virgil and others, sailed from Troy after the famous siege, stopped briefly for an affair with Queen Dido at Carthage in North Africa, and finally, spurred by the call of duty, abandoned Dido and set off to found the city of Rome as a new Troy. Dido, it is said, was so mortified by his departure that she had herself burned to death in a gruesome, exhibitionist ritual.

Changing calendars is easy enough to do; everyone likes to chip in with ideas on how it should be done. The problem, as we shall see, always lies in making your changes stick. Only two of our 12 months are still named after emperors. Some countries have avoided the Roman influence altogether. The Dutch, for instance, have always called November *Slaght-maand*, which is similar to the Viking *Blot-monath*. They mean respectively 'slaughter month' and 'blood month', and refer to the time of the year when the animals were traditionally butchered and salted or smoked for winter preservation. The Saxon *Wind-monath* and the French Republican *Brumaire* describe November as the month of wind and fog. October was referred to as the 'wine' or 'vintage' month in several languages, and December, the winter month, was to the Revolutionary French the onomatopoeic *Frimaire* or hoarfrost month.

The Jewish calendar was, since the time of the Babylonian exile, influenced by the lunar cycles. It was composed of a 354-day year, plus an intercalated month every three years. Long before that, it is alleged that the Jewish calendar was solar-based, holding to a year of 364 days, or precisely 52 weeks, 12 months of 30 days each, plus four intercalated days. This neat system, invented, it is said, by the visionary prophet Enoch, enabled the first day of the year, as well as the first day of each of the four seasons, to fall on the same day (i.e. a Wednesday). Likewise, under Enoch's system the Passover and other Jewish feast days would fall on the same day of the week each year. Perhaps it is for this reason that Enoch who, in the Bible, 'walked with God' was rewarded for his efforts with a lifespan of 365 years – what a record!

The Book of Jubilees, an apocalyptic text from 150 BC which does not appear in the Bible, suggests a quarrel between those Jews who operate a lunar-based calendar and those who follow the sacred plan of Enoch. At the Creation we are told: 'God appointed the sun to be a great sign on the earth for days and for Sabbaths and for months and for feasts and for years.' The angel Uriel, God's messenger, tells Moses that the Jews must follow the solar calendar, for if they do not 'then they will disturb all their seasons and years will be dislodged'. Uriel prophesies that this will indeed happen. Holy days will be confused with 'unclean' days when the true times of Sabbaths and other holy feasts are forgotten. According to the teaching of Jubilees, God burst down through the clouds and landed, with 10,000 angels, on top of Mount Sinai to administer the Last Judgement. In the kingdom of heaven it is the solar year that shall prevail. Jews who have been operating by the lunar calendar will not be admitted through those pearly gates. On the whole, Jews have tended to ignore Uriel's warning from the Book of Jubilees and carried on with their lunar-solar calendar, which was an ancestor of the Jewish calendar that is still in use, by the orthodox community, today.

The earliest history of the Roman calendar is obscure. Like most measuring systems it must have been, at least in part, inherited from somewhere else. The distant origins are most probably Egyptian. We know that the Egyptians, despite their many calendric innovations, were indebted in the most part to the Babylonians and that the Babylonians, in turn, evolved their calendars from the earlier Mesopotamians. If we search for the origins of the Mesopotamian calendar we get back, before you know it, to our old friends the Sumerians, who were known to base their calendar on a primitive but effective bi-seasonal, lunar year.

The Romans, too proud to admit that they had drawn from any other culture, encouraged the myth that the first Roman calendar was designed by Romulus, seven centuries before Christ. Whether he did or did not have a hand in it, Romulus could at least claim to be the first individual whose name is associated with the evolving life of the calendar. Unfortunately he never existed, or if he did he was almost certainly not the son of the vestal virgin Rhea and the brutal god Mars. This, we suspect, was a bit of wishful thinking by the Romans hoping to connect their city with the god of war. Romulus and his twin brother Remus floated down the Tiber in a pigs' trough instead of drowning as the jealous king Amulius had hoped. The infant twins were saved by a woodpecker and suckled by a wolf (an act to which many fine Roman statues have attested), later finding the strength to depose Amulius and restore his brother to the throne. Romulus soon came to power himself, killed his brother, raped the Sabine women, then used them to persuade their menfolk (the Sabine men) not to seek revenge. Little more is known of Romulus except that, after a long and fruitful rule, he disappeared one night in a storm and never came back. The Romans, impressed by this, deified him as the god Quirinius.

Romulus's calendar design began the year with the month of March. His scheme incorporated six months of 30 and four months of 31 days, making a total year of 304 days. Clearly this did not work in so far as aligning itself to the seasons was concerned. Ovid, a Roman poet living 700 years after Romulus, said the king 'was better versed in swords than stars', showing that, while the Romans were keen to have Romulus as the father of their city and their calendar, they were not altogether happy with his work.

Romulus was presumably working on some form of decimal numbering system, which separates him and the Romans from their Mediterranean neighbours, the Egyptians and the Greeks. They used compound systems based on decimal, duodecimal and remnants of the Sumerian sexagesimal systems. Romulus's army and his senate were comprised of multiples of 10. His 10-month calendar was naive but not unexpected. What was strange about it were the names he gave to each month: *Martis, Aprilis, Maius, Junius, Quintilis, Sextilis, September, October,*

November and *December*. Why the first four should have exotic fancy titles and the proceeding six be given numbers for names has never been adequately explained. Is it possible that before this there were only four months in the year? Or were *Martis*, *Aprilis*, *Maius* and *Junius* names previously given to seasons and subsequently allotted to months in the new 10-month calendar? We do not know.

What is interesting, though, is that Numa Pompilius (*c* 714–675 BC), supposedly Romulus's successor in Rome, so quickly realised that Romulus's system did not work. With only 304 days in the calendar year, the months will shift by 61 days a year against the seasons. The third day of June, for instance, might find itself in high summer in one year and in early autumn the next, which would make the calendar useful only for counting small groups of days, and pretty useless for long-term planning. In response, Numa Pompilius is said to have extended the year by 50 days, adding January and February and bringing the total number of months up to 12, as it is today. But first of all, Pompilius subtracted one day from each of the six months containing 30 days. This was doubly useful, because the Romans suffered, as we have seen, from bad anxiety attacks about even numbers.

The number one stood for unity; two for diversity and therefore disorder and chaos. Anything that divided by two must (to the worried Roman) be chaotic, suspicious and scary. For this reason they dedicated the second month to Pluto (ruler of the infernal underworld) and the second day to Manes (spirit of the dead).

By thinking sensitively about even numbers, Pompilius was able to turn six months with an even number of days into six months with an odd number of days, while, at the same time, freeing up six extra days which he could then distribute among his two new months, January and February. But the problem for the king was not yet over.

Dividing the 56 days between January and February would create two new months of 28 days each – even numbers! Oh help! Adding one extra day to each month would not restore confidence among his people – for they would surely spot that the year was now 356 days long and be scared out of their wits by that awful even number. The solution had to be a compromise. He added one day to January, giving it an odd number of days, and making the year 355 days long: that made for two areas of satisfaction, leaving only February to blub about. He assembled the people and explained to them that, since this month was devoted to infernal gods, the even number of days might be judged appropriate after all. The people sighed a long sigh of relief and, that night, slept like logs in their *cubicula*.

But there was *still* a problem. The 355-day year, if it was not working as an adequate representation of a solar year, might at least have been expected to serve as a reasonable key to the lunar year. Not so: it worked out to be 16 hours, or 0.66 of a day, longer than a lunar year, putting the calendar out by two days every three years against the actual orbits of the moon.

Quintus Ennius, poet and father of Latin literature, mentions an eclipse of the sun which was supposed to have occurred on 5 June 355 BC. If the Roman calendar had been in tune with the phases of the moon, this would have been impossible, for on the fifth day of any month the moon ought to have been in its first phase (i.e. a new moon) and not, therefore, in any position to eclipse the sun. The evidence shows that the calendar had already drifted significantly from the true celestial lunar month.

Pompilius's system lasted for only another three generations, for Tarquinius Priscus (617–578 BC), known as the fifth king of Rome, famous for instituting the *ludi romani* (Roman games) felt a need to tinker. It was Tarquinius's grandson, Tarquinius Sextus, who committed the famous rape of Lucretia which led to the expulsion of the Tarquin family from Rome. Lucretia was the wife of Tarquinius Sextus's friend and political rival, the general Collatinus. Famed for her faithfulness at a time when most generals' wives were taking advantage of their husbands' long absences from home, Lucretia was visited by Tarquinius Sextus one windy night, on the pretext that his horse had a lame leg. The upshot, as we all know, was rape – motivated by Tarquinius's fear that Collatinus would win over political supporters from him on the basis of his wife's good name. It was a deliberate attempt to sully her reputation.

As it happens, the Tarquin family were considered upstarts by the aristocracy of Rome, who were looking for any excuse to depose them. The rape of Lucretia provided the aristocrats with all the excuse they needed. Tarquinius Priscus, the rapist's grandfather, had been instrumental in changing the first month of the year from March to January, but the effect of the rape in the family was such that the right-thinking people of Rome attempted to erase all record and memory of the odious Tarquins from their city history. This meant that January was put back to where it always was (somewhere in the middle) and March returned to its original spot at the beginning of the Roman calendar.

In the 500 years from the reign of Tarquinius Priscus to that of Julius Caesar, the calendar fell under the control of the pontifices, a corrupt board of powerful priests. But they were more than priests – a far remove from the cosy tea-and-cakes vicar that we have come to associate with the cloth today. In Rome the pontifices controlled parts of the legislature and were supported by the spoils of taxation. 'Pontifex' means 'bridge-builder' because one of their special tasks was to administer the *jus divinum* ('divine law') which was a part of the civil law that regulated the relations of the community with the state-approved deities. In this capacity the pontifices held enormous sway and could juggle things in any way that suited them without being under any obligation to explain to anyone the reason for their edicts. At the time of Julius Caesar there were 16 of them, living in the *Capitol Collegium* surrounded by vestal virgins, *flamines* and other useful assistants. They regulated special ceremonies (sacrifices in response to plagues or thunderstorms); they also controlled marriages and wills, and were in sole possession of the state archives. Chief among their tasks was the regulation of the calendar.

The result was chaotic, for the pontifices irregularly added days to suit themselves – to recoup a loan early, to shorten an elected senate's term of office, or to avoid paying a vexatious bill. Their wayward behaviour was most notable with regard to the collection of taxes. In those days tax collectors were not white-collar clerics with 2.2 kids, a hatchback and a suburban house with a containable mortgage. Roman tax collectors were powerful enough to make considerable fortunes in the execution of their duties and, as a result, were loathed by the people with far more vengeance than we nowadays feel for the modern, monochrome office clerk. Suspicions arose that the pontifices and the tax collectors went about with their hands in each others' pockets. An additional day or two, plopped by a non-chalant pontifex into the Roman calendar, could mean colossal extra profits for the tax-man, while the pontifex himself was conveniently protected from having to reveal anything about how or why he reached his decision to add the extra days.

This is exactly what happened.

The calendar year, as we have seen, was at that stage 355 days long. March, May, July and October had 31 days, February consisted of 28 and the rest of 29. The pontifices were only too aware that this 355-day calendar year was not in line with the solar year, and the seasons were consequently slipping against the calendar. It was their job, therefore, to add extra days in order to restore the balance.

Intercalans or *mercedonii* were supposed to repair the damage by providing an early form of leap-year system. They were groups of 22 or 23 days (depending on the slip in the lunar calendar) inserted, supposedly into the month of February, once every other year. The word *mercedonius* comes from *merces*, meaning 'wages' (hence 'mercenary soldiers'), and it was at this time that civil workers expected to be paid (unless, of course, the pontifices were their masters). Over four years the effect of the additional *intercalans* days (assuming they were properly administered by the pontifices) should have created a period of 1,465 days: divide that by four and you get an average year of 366¼ days, just long of the mean solar year by one day.

But from the start there were problems. For no known reason

the *mercedonius* was slotted into February after the twenty-third. This left 24–28 February dangling. Were they to follow the *mercedonius* or not? If not, should they be eradicated from the calendar during a *mercedonius* year? It was in this area that the pontifices profited most from the maximum confusion by allowing the *mercedonius* to be either 28 or 27 days long, or if the last five days of February were removed, either 23 or 22 days long. The pontifices could keep everyone guessing right up to the last minute as to which way they would jump.

This was clearly not a situation that Julius Caesar, emperor and supreme commander, could tolerate for long. Behind his need to control the calendar was a power battle between the *Collegium* and the imperial office. In 63 BC Caesar became *Pontifex Maximus*, head priest of the *Collegium*, a post he held to the end of his life. After the death of the subsequent holder of that office, Pontifex Maximus Lepidus, in 12 BC, the position went to Emperor Augustus and, from there on, remained a job taken, on principle, by the reigning emperor. As *Pontifex Maximus*, Caesar was able to devise a self-regulating system for the calendar which effectively put an end to the corrupting powers of the pontifices to control the whens, the wheres and the ifs of the *mercedonii*.

Caesar, like his successor as emperor, Augustus, was something of a narcissist. He was less attractive to look at than Augustus (no historian ever described Caesar as handsome), yet his power and remarkable energy proved alluring to many. He enjoyed a homosexual liaison with King Nicomedes of Bithynia and, more famously (and more controversially), had a heterosexual affair with Queen Cleopatra of Egypt. His nose was aquiline, his chin narrow, his mouth mean. The mistakes he made were few, but historians believe that his undoing was caused by the magnanimity he bestowed on his defeated Roman adversaries. Towards barbarians Caesar was merciless. '*Veni, vidi, vici*,' his famous saying goes ('I came, I saw, I conquered'). But to his Roman adversaries, who were legion, he was noticeably merciful in victory. Among his assassins were Gaius Cassius Longinus and Marcus Junius Brutus, both former enemies whom Caesar had forgiven and appointed to high office. *'Et tu, Brute'* ('You too, Brutus') was Caesar's baleful exclamation on discovering that his former enemy and now close friend was one of these men.

In Shakespeare's play *Julius Caesar*, Cassius, who hates Caesar, is nevertheless forced to admit:

> . . . he doth bestride the narrow world
> Like a Colossus, and we petty men
> Walk under his huge legs, and peep about
> To find ourselves dishonourable graves.

To Mark Antony also he was '. . . the noblest man that ever lived in the tide of times.'

Given that he died at the age of 44, Caesar's achievements were little short of miraculous, for on top of all the military campaigns, the restructuring of civic affairs, the rooting out of corruption, the squashing of civil insurrection, the chasing of his arch-rival Pompey up and down the empire, and the endless marching hither and thither, not to mention the judicious politicking that needed to go on in Rome, amid all this Caesar also found time to write books, many of them: annals of his campaigns, letters, speeches, addresses of all sorts, funeral orations, some of which have survived. His work on the calendar, which brings him into our story, was put into effect two years before his death.

As far as calendar reform was concerned, Caesar needed to gain control of the pontifices who were running amok and stop them from dropping their *mercedonii* on the calendar whenever and however they felt fit. The problem he faced was how to create a calendar that would run on its own, with no need for priests or tax-men to make irregular additions and subtractions, as suited their book. In his quest he was helped by Sosigenes, an astronomer from Alexandria. Do not confuse him with Sosogenes the Peripatetic, tutor to the philosopher Alexander of Aphrodisias – they are not the same person. Egypt was a conquered nation, but the Romans still respected its people and its ancient tradition of scholarship. In particular, the Egyptians were revered for their unparalleled knowledge of mathematics and astronomy, and, at the time of Caesar, Sosigenes was thought to be one of the wisest men alive. Only fragments of his work survive. His book *Revolving Spheres*, which was known to scholars across the empire, no longer exists; all we know of it now is that it contained the proposition that Mercury revolved around the sun.

Sosigenes's suggestion to Caesar was not altogether original. He referred his emperor back to a solution that had been proposed for a reform of the Egyptian calendar by Ptolemy III some 100 years earlier.

Ptolemy was essentially a Greek – being the grandson of Ptolemy I Soter (367–283 BC), one of Alexander the Great's most brilliant military commanders who acceded to the throne of Egypt after Alexander's death. The Egyptian old guard were naturally suspicious of the Ptolemaic dynasty for this reason. The first three or four Ptolemaic kings proved to be able and trustworthy rulers. Ptolemy III's father was Ptolemy II Philadelphus, who founded the great museum and library at Alexandria into which he deposited thousands of valuable manuscripts of Greek literature and established the city as the ancient world's capital of scholarship. Ptolemy III was known as Ptolemy Eurgetes ('the benefactor') and it was he who proposed the leap year to the Egyptians. As in Rome, tinkering with the calendar in Egypt was considered the divine right of priests. The Egyptians had for thousands of years been using a 365-day calendar year, which was explained to its people as a 360-day year with five extra-special festival days thrown in for religious reasons by the priests. But 365 days did not, as everyone knew, make a full solar year. The earth takes $365\frac{1}{4}$ days to get around the sun (365.242199 days if you wish to be pedantic), so to ignore that quarter of a day would result in a calendar slippage of one full day every four years.

Ptolemy could not have suggested a more amenable or a more sensible solution. All you need to do, he told his people, is to add an extra festival day once every four years and the calendar will be in balance with the solar year. And so started the leap-year system which, once every fourth February, is still in use today. Well – not quite.

Egyptian priests were appalled by Ptolemy's sacrilegious suggestion and many of them flew into uncontrollable fits of rage. The five sacred Epagomina festival days were not, under any circumstances, to be turned into six. These were ancient laws that had been in place from time immemorial, and it was not for some upstart Greek to come and tell them to change them. The priests won the day, and Ptolemy's suggestion was shelved.

Julius Caesar was not in the same position as Ptolemy III had been. He did not have holy numbers to fret about; to him it could not matter less if the year was seen as 360 days plus five intercalated festival days, or simply as 365 days. He did not have to worry about angry priests railing at a break with tradition, since he himself was now *Pontifex Maximus*. He saw the need to account for the extra quarter of a day, and the leap-year system suggested to him by Sosigenes was clearly the solution he needed.

The year 46 BC thus came to be known as the Year of Confusion, for it was this year into which Caesar piled his reforms and emancipated the calendar from the office of pontifex. Into 46 BC Caesar put a 23-day *mercedonius* and added two further months of 67 days between them, converting the Year of Confusion from a 12-month, 355-day year into a 15-month, 445-day year (i.e. 355 plus 23 plus 67 days). This excited certain of Caesar's provincial governors, who instantly issued demands for two months' extra tax from the people, resulting, as the name so succinctly informs us, in a year of utter chaos and administrative hell.

Once the Year of Confusion was over, however, Caesar's new system, known as the Julian calendar, ticked over (with a few hiccups, it must be said) for the next 1,600 years, and it is this ancient system that still forms the basis of the calendar in use today. The following year, 45 BC, was organised to a system easily identifiable as the origins of our modern calendar. January was re-established as the first month of the year – the disgrace of the Tarquins having been put to one side – and contained

31 days. February had 29 days and 30 in a leap year. March, April, May, June, *Quinctilis* (which promptly changed its name to July), *Sextilis*, September, October, November and December alternated between 30 and 31 days each. The only anomaly was the naming of the last five months (*Sextilis*, September, October, November, December) which, when the first month of the year was changed to January, continued to be counted as though the first month were still March.

If the pontifices had only obeyed instructions all would have been more or less fine. Unfortunately, Caesar was dead within a year, struck down by his enemies in the senate, and the pontifices, either failing to understand his instructions, or deliberately misinterpreting them, instituted a leap year every third instead of every fourth year. Caesar was not around to admonish them. The year 8 BC was consequently three days late in arriving and something had to be done about it. Augustus, by then emperor as well as *Pontifex Maximus*, took action by suspending leap years until AD 8, ensuring that the three erroneously gained days were put back to right. Leap years have run consistently from AD 8 to the present, which is why all leap years have occurred in year numbers which are divisible by four ever since.

From the time of Augustus there were irregular attempts to change the names of the months. We have already mentioned those vainglorious emperors who hoped to enter their own names into the calendrical pantheon but failed. The Frankish Emperor Charlemagne (from whom three-quarters of the world's genealogists are wont to claim descent) made a similar attempt, not in order to glorify himself but to sever symbolically the connection with Roman imperialism. Like Ptolemy II, Charlemagne was an intellectual, disgusted by the stupidity and moronic abilities of his own subjects. In earnest, he set about the task of educating all the people of his colossal empire. 'Let them learn psalms, notes, chants, calculation and grammar in every monastery and every bishop's house and let those who can, teach' – a curious mirror of George Bernard Shaw's observation: 'He who can, does. He who cannot, teaches.' Charlemagne would not have cared for that. People who knew the emperor described a cheerful, friendly countenance. He was widely regarded as equable and fair-minded in judicial dealings and was consequently popular among his subjects. If he had an odd side it usually concerned his daughters. He was obsessive about them, scarcely letting them out of his sight. Nor would he allow them to marry. He was therefore obliged to accept the consequences. Bertha had two illegitimate sons by the fat-faced Abbot Englebert of St Riquier. Charlemagne's court was reputed to be a centre of very loose life. Einhard, in his *Vita Caroli*, which contains the most important extant first-hand account of Charlemagne, seems to hover precariously between insult and compliment in his description of the Holy Emperor's looks:

> Seated or standing he made a dignified and stately impression even though he had a thick, short neck and a belly that protruded somewhat; but this was hidden by the good proportions of the rest of his figure. He strode with firm step and held himself like a man; he spoke with a higher voice than one would have expected of someone of his build. He enjoyed good health except for being repeatedly plagued by fevers four years before his death. Towards the end he dragged one foot.

One can but imagine this high-voiced, pot-bellied, cheerful-faced potentate and the pleasure he must have gleaned in renaming the months:

'How about *Wintarmanoth* for January?'

'Sounds good, my lord!'

'Any objections to *Brachmanoth* for say . . . June?'

'Indeed I have none, my lord.'

'Now I know you keep saying that *Aranmanoth* is a good one for February, but I'm still convinced that *Hornung* would be better. You see, *Aranmanoth* is an Augusty sort of name. No, it's got to be *Hornung* – *Hornung, Hornung, Hornung*. It's so very February, don't you agree?'

'Indeed, my lord, so very February.'

Charlemagne's hopelessly unsnappy month names, which did not even make it on to his own gravestone (the inscription at Aachen reads 'He died at the age of seventy, the year of our Lord 814, the 7th Indiction, on the 28th day of January'), offer an interesting insight into the French mind and found a strange parallel in France nearly 1,000 years later.

At the end of the eighteenth century, the French, in the young buck spirit of a people on the cliff edge of revolution, decided that the names of the calendar months smacked of Papal authority and Roman imperialism. After the storming of the Bastille in July 1789, the French populace was screaming for reform of more or less anything and everything. High on the list was a desire to institute a new calendar to start 'from the first year of liberty'. Philippe-François Nazaire Fabre d'Eglantine was the flowery name of the man chosen for the job of renaming all the days and the months. If we think that Charlemagne enjoyed saying *Brachmanoth* and *Hornung* to himself, Philippe Fabre d'Eglantine must surely have been in second heaven! Not only, as we have already seen, was he allowed to give fancy titles to all 360 days of the year, but he was also granted *carte blanche* to rename the months as well. All his recommendations were accepted without delay and were officially ratified as the Calendar of Reason on 5 October 1792. The names of the months chosen by M. Fabre d'Eglantine were as follows:

Vendémiaire ('vintage'): 22 September – 21 October
Brumaire ('mist'): 22 October – 20 November
Frimaire ('frost'): 21 November – 20 December
Nivose ('snow'): 21 December – 19 January
Pluviose ('rain'): 20 January – 18 February
Ventose ('wind'): 19 February – 20 March
Germinal ('seedtime'): 21 March – 19 April
Floréal ('blossom'): 20 April – 19 May
Prairial ('meadow'): 20 May – 18 June

Messidor ('harvest'): 19 June – 18 July
Thermidor ('heat'): 19 July – 17 August
Fructidor ('fruits'): 18 August – 16 September

Eglantine was one of the cream puffs of the French Revolution. A lofty portrait by Jean-Baptiste Greuze, which hangs in the Louvre, leaves one in no doubt that he was a mincing, prissy individual with excessive self-regard and a moral righteousness that must have nauseated his contemporaries. He added the 'd'Eglantine' element to his original surname of 'Fabre' in an effort to aggrandise himself (a strange thing for a republican revolutionary to do!), claiming that the reason he chose that particular name was because he had received a golden eglantine (the equivalent of a literary Oscar) in a recent poetry competition. This was a lie. He had never won a golden eglantine in his entire life and, to this day, his most famous poem (if one can call it that) is the cutesy little nursery rhyme '*Il pleut, il pleut, bergère*' ('It's raining, it's raining, shepherdess'). In 1793 he voted for the execution of Louis XVI (a vote which was won by a margin of 380 to 334). But the joy and smug self-satisfaction which Eglantine must have felt at having a democratic hand in the execution of his sovereign was short-lived. For in the following year Eglantine had his own head chopped off at the instigation, it is thought, of Maximilien Robespierre, the Franco-Irish political leader of the Revolution. But these were bloody times for all. Three months after Eglantine's execution, Robespierre himself had his jaw shot off following an attempt to resist arrest by the National Guard. The next day he was placed, without trial, on the scaffold, where his own jawless head was promptly chopped off.

When Fabre d'Eglantine was marched to the guillotine he must have carried with him a certain pride in his life's achievement. After all, he had succeeded in single-handedly renaming all the days and the months of the calendar. He could comfort himself, in his final hours, with these last warming thoughts: 'At least my passing shall be for ever recorded as having occurred on the Bumblebee of Seedtime in the Second Year of Freedom.'

Unfortunately, Fabre d'Eglantine's bumblebees and seedtimes hardly made it beyond his funeral, for although the French begrudgingly accepted the mechanics of his new calendar when dealing with internal matters, they, and all foreigners, were exasperated with the effects it had on international trade relations. French months were constantly changing their relationship to months in the calendars of neighbouring countries, which made for nothing but cross-border chaos. By September 1805 the scheme had been virtually abandoned. The official death knell was sounded on 1 January 1806, when the French nation finally returned – tails between their legs – to the fold of an established system which had been instituted by Pope Gregory XIII 250 years earlier.

CHAPTER VIII

ANNI
Years

The period immediately after AD 8 was peaceful for the calendar, but a long-term problem was brewing. Mean solar years were not exactly 365¼ days long, so the year was still refusing to harmonise precisely with the calendar. The difference may have been small, and perhaps the Romans were aware of it all along. If they were, they knew that the effects would never seriously impinge in their own lifetimes. It was an error amounting to 11 minutes and 14 seconds per year, enough to put the calendar out by one and a half days in two centuries, or a week in 1,000 years. Unfortunately, what happens with time is that it never stops rolling along, and, eventually, the ever-widening discrepancy between the Julian calendar year and the solar year had to be faced up to.

During the Dark Ages the slip was considered and pondered over by many learned men including the Greek astronomer Claudius Ptolemy, the Indian Aryabhata and the venerable English monk, Adam Bede. These were the famous names among a host of obscurer monks and dusty theologians who all tried to find different solutions with which to right the error.

An English friar, Roger Bacon (1214–92) wrote, at the behest of Pope Clement IV, his *Opus Maius*, which, in one part, set out to tackle the issue. Bacon refers in his work to the error of the Julian calendar, angrily denouncing it as 'borne of ignorance and negligence contemptible in the sight of God and of holy men'. To many milder-mannered people, Bacon was simply mad. His aggressive stance towards any who disagreed with his often fearsomely avant-garde views landed him in the soup on more than one occasion. Although he was, towards the end of his life, a Franciscan, his faith could not save him from imprisonment by the brotherhood on a charge of incorporating 'suspected novelties' into his teachings. The assumption that Bacon was mad is given strength by the strange story of the brazen head. Perhaps he had been reading too much fantasy fiction; he was known to be an avid collector of 'secret' books and spent large sums of money building up his library. In one of his books he probably read the legend of the magic Eastern head – a giant, mythological statue made of glowing brass that would answer any question put to it, on any topic, past, present or future. For a scientist with an eager, enquiring mind, the magic head must have been the most alluring toy imaginable, but to more realistic minds (and it must be said there were not many of those knocking around in the thirteenth century) it remained 'such stuff as dreams are made on'. Bacon, undeterred by the forces of reason, such as they were, set about making a magic head at his home in the west of England, promising himself that if it spoke he would succeed in all his projects; if it remained dumb, he would fail.

Though a busy man, Bacon kept a permanent vigil on his brazen head and at night he set a servant to watch over it. One evening, when Bacon was tucked up in bed, the head spoke. It spoke to his servant, Miles, articulating three phrases, each separated by an interval of half an hour: 'Time is' . . . 'Time was' . . . 'Time's past'. What a cryptic allusion that was. 'Quoth he,' as Samuel Butler said, 'My head's not made of brass/ As Friar Bacon's noddle was.'

Insane or not, Bacon was without doubt one of the most fascinating figures of English medieval history. He studied flying and built a machine with flapping wings, he was a pioneer in optics, he forged himself a pair of useful, functioning spectacles, he designed a mechanically powered car and a mechanical boat, and he made a camera obscura for observing eclipses. On top of all this, he wrote long, waffling encyclopedias full to the brim of obscure, yet often fascinating scientific, mathematical and philosophical thought. Unfortunately, none of his ideas took a hold during his lifetime, and although he is now regarded as one of the most visionary minds of his age, to his contemporaries he was little more than a cantankerous, self-opinionated loony. As far as we know, his solution to the calendar problem was never even read by the Vatican; certainly nothing was done about it until 300 years after his death.

The reason that the Church became so immersed in the history of the calendar has its roots in the Bible and our ancient scriptures. In order to celebrate the anniversaries of Christ's birth or his ascension, it was important to know when these events were historically supposed to have occurred. If Christmas is supposed to be a winter celebration (if, for instance, the Bible had said that Jesus was born in a snowstorm typical for the time of year), then neither the Pope in Rome nor any other Christian leaders would wish to see subsequent anniversaries of this holy event slipping (by fault of human error in the calendar) into summer or autumn. We will come to Christmas later on. Easter, meanwhile, the anniversary of Christ's resurrection, has always provided the thorniest problem.

There are various means by which the Church has attempted to date the crucifixion. We know, for instance, as an historical fact, that Pontius Pilate was Roman Governor of Judaea at the time of Jesus's death, which, on the evidence of two early commentators, Tacitus and Josephus, allows us to narrow the year of the crucifixion to somewhere between AD 27 and AD 36. Now if the Last Supper was, as Jewish scholars believe, the feast of the Passover, then it would have taken place on the afternoon of the fourteenth day of *Nisan* (the first month of the Jewish religious year). This is clear enough, but unfortunately a squabble has been generated as to when Jesus was crucified in relation to the Passover/Last Supper. St John's Gospel says one thing while the Gospels of Matthew, Mark and Luke say another. What is accepted with unanimity by all the Gospels is that the crucifixion occurred on the Jewish Day of Preparation, which is the day before the Sabbath. In other words the crucifixion must have fallen on a Friday. That Jesus rose from the dead 'on the third day' is, to some people, an indisputable fact which puts Easter Day on to the Sunday following the crucifixion and leaves scholars and astronomers to bicker about the exact date of the crucifixion itself. Was it Friday 7 April AD 30 or Friday 3 April AD 33? Given that Jesus is believed to have been between 33 and 34 years old when he was nailed to the cross, theologians quickly and without fuss have settled for Friday 3 April AD 33 as the true date of the crucifixion. Wrong!

Dionysius Exiguus (a character who shall be given fuller treatment later in our narrative) invented, it is said, the BC/AD dating system, and in so doing made a possible mistake about the date of Jesus's birth. If the early scriptures are to be believed, Jesus was born in the lifetime of King Herod the Great, who died in 4 BC, which would have made Jesus 33 or 34 years old at the time of his crucifixion cited on 7 April AD 30.

The upshot of several hundred years of bitter quarrelling between two rival factions of theologians, calling themselves the Quartodecimans and the Quintadecimans, meant that the date for Easter was finally decreed as falling (in any given year) on the 'first Sunday after the full moon occurring on or after the vernal equinox'. This means that Easter can, and does, fall anywhere between 22 March and 25 April. It also means that the calculation of Easter is dependent on observation of both the moon (since it requires a full moon) and the sun (requiring observation of the vernal equinox). Now a slipping calendar can, in theory, disperse Easter to any distant corner of the civil calendar, and since Christianity is based on supposedly true historical events it is of the utmost importance to the authority of the Papacy, and to other Christian church leaders, that the dates of the Christian calendar have an aura, at least, of verisimilitude about them. Which explains why Pope Clement IV was corresponding with Roger Bacon about the mistakes in the calendar, why the Venerable Bede was involved, and why Bacon himself described mistakes in the Julian calendar as 'ignorance and negligence contemptible in the sight of God and of holy men'.

It also might help to explain why Pope Gregory XIII took it upon himself to reform the calendar as a matter of urgency in 1582.

The Julian calendar, as has been explained, was too long – not by much, but enough to cause a problem. It should have been 365.242199 days long, but with Caesar's leap-year adjustments making it 365.25 days, the annual error of 11 minutes and 14 seconds had, by 1545, caused the vernal equinox (that all-important yardstick for determining Easter) to have moved 10 days off from its original calendar date.

Action had to be taken. The festival of Easter was turning into a mockery. It was now being celebrated on a date that was clearly not the true anniversary of the crucifixion. But neither the sensitive and talented Pope Paul III nor any of his able-minded aides, try as they might, were able to work out a solution. Paul III was followed in the papacy by the dull and ignoble Julius III, who did nothing whatsoever to address the issue. Julius III's successor, Marcellus II, died almost the second he had taken office. Paul IV, who took his place in 1555, is best known for insulting Elizabeth I, Queen of England, by telling her ambassador to the Holy See that Her Majesty was a bastard. He passed the problem of the calendar on to his successor, Pius IV, a member of the all-powerful Medici dynasty, who became Pope in 1566, founded the Vatican press and stuffed all things to do with the calendar into his bottom drawer where . . . they were once again ignored by his pious and energetic successor, Michael Ghisteri, known to us now as Saint Pius V. When Cardinal Ugo Buoncompagno was elected to the papacy in the name of Pope Gregory XIII, excuses were running out and Christian monks and scholars throughout the Catholic world were urging the Vatican to sort the matter out.

Gregory XIII took it in hand and finding, on his arrival as Pope, a number of conflicting proposals for calendar reform, issued a bull (a papal edict) drawn up by the astronomer Christopher Clavius on the advice of a little-known physician from southern Italy, Dr Antonio Lilio.

Antonio was the brother of Luigi, also a physician, whose Latinate name is sometimes seen as Aloysius Lilius. The surviving biography of Luigi is not a full one. He is believed to have been the son of an average middle-class couple and was born, it is thought, around 1510 at Ciro in southern Italy. He studied astronomy and medicine at Naples, later moving to Verona (what he did there history does not relate) and spending the years until his retirement immersed in pedagogic activities at the University of Perugia. At the end of his career he retired to Ciro, where he is believed to have died from a pancreatic complaint in 1576. The reason we have sidetracked from Dr Antonio Lilio to his seemingly dull and little-known brother Luigi is because it was Luigi, not his brother, who single-handedly invented the solution to the problems of Caesar's Julian calendar, known as the Gregorian Reform, which is still in use today. We suspect that Luigi must have been working on the calendar close to the time of his death, which would explain why he never managed to submit it to the Pope himself. It was his brother Antonio who took Luigi's proposals to the Vatican. By the time Gregory XIII had been elected in 1572, the Lilio brothers' proposal had wormed its way to the top of the pile. Antonio accepted an advisory role and worked closely with Clavius towards their final solution.

Christopher Clavius (1538–1612) was a Jesuit priest, which means that he was clever, single-minded and probably very devious. Jesuits are members of the Society of Jesus, 'founded in 1534 by St Ignatius Loyola (alias Iñigo López de Recalde). Recalde was a soldier and an aristocrat with a fine *castillo* in the Basque province of Guipúzcoa. But in a moment of headstrong religious zeal (born from a book about Jesus that he had read while convalescing from a broken leg) he cast aside his life of luxury and set himself up as an intellectual ascetic, focusing, with St Francis Xavier and other friends, on a new order he described quaintly as a 'little battalion of Jesus'.

Love, in point of fact, was not a key word to the Jesuits. They were the crack troops of the Catholic Church, a highly disciplined, organised and secretive team. Their mission was to infiltrate the highest echelons of society in the hope that their message would percolate downwards to the ordinary people. To this end, the Jesuits impressed themselves upon landowners, aristocrats, kings, emperors and potentates, converting them all to the strictest doctrines of the Roman Church and encouraging them in an attitude of zero tolerance towards the growing protestantism of the Reformation. As time went on, the Society of Jesus expanded its numbers (there were 22,589 members by 1749), and by the end of the century, Jesuit missionaries had seated themselves at the right hand of many of the world's most powerful leaders: the tsars of Russia, the emperors of Japan, the maharajas of India and the kings of France and Brazil.

Unfortunately for the Jesuits, however, so much power allied to so much secrecy was a recipe for disaster which led inevitably to a festering contempt for the order right across Europe.

Widespread revulsion against the Jesuits quickly spread to all corners of the globe. The order was eventually expelled from France in 1594, from England in 1579, from Venice in 1607, from Spain in 1767 and from Naples in 1768. By 1773 the Jesuit order had gained such a rotten reputation that the Pope himself (Pope Clement XIV) felt the whole Roman Church to be besmirched by the bad name of the Jesuits and promptly banned the whole lot of them. While the order still exists, the word Jesuit or Jesuitical is applied in a derogatory sense today, having taken on a secondary meaning of 'deceiver' and 'prevaricator', one who 'lies like truth', who 'keeps the word of promise to our ear, and breaks it to our hope'.

That Clavius was a Jesuit is relevant to the task he was chosen to perform. Gregory XIII knew that, even with the greatest powers allotted to him as Pope, he was going to have an uphill task persuading the world to accept his reforms. Like a whip in the House of Commons, Clavius would creep through the corridors of the Vatican unctuously whispering the words of calendar reform. He worked assiduously on the project before and after the reforms were enacted in 1582. By 1606 he was able to pose – plump, whiskered and self-satisfied – for a portrait which survives in the Vatican today. He had earned himself the sobriquet 'Euclid of our times'; such a revered and clever man was he that even the great scientist Galileo came to him for support and advice. Being a Jesuit, though, he was blinkered in certain respects. While supporting Galileo in his ground-breaking work with the newfangled telescope, he refused to accept the Copernican view that the sun, and not the earth, was at the centre of our solar system. To accept that would be to renounce the teachings of the founding father of his order, St Ignatius.

Clavius was undoubtedly the force behind much of Pope Gregory's notorious censorship of books, which included a ban on Copernicus's seminal work of astronomy, *De Revolutionibus Orbium Coelestium* (*On the Revolutions of the Celestial Spheres*) of 1543. It may also have been Clavius who suggested to his newly elected Holiness that the appropriate Vatican response to a recent, bloody massacre of 800 protestant Huguenots in Paris should be a celebratory Te Deum (a hymn of thanksgiving) to be sung with full voice and hearts uplifted in St Peter's Basilica, 1582. The Pope compounded this gloat by minting a special limited-edition medal which depicted the massacre in glittering metal; it was put on sale for collectors and made available at all good Vatican shops.

Clavius's name as a scientist survives as scarcely a footnote – he backed the wrong horse by rejecting Copernicus's correct analysis of the solar system, and that sealed his posthumous reputation for the worse. But his work on the promulgation of Lilios's reforms will always permit the name of Clavius to creep into books as long as the history of calendars remains of sufficient interest to warrant a pressing.

In his satire *Ignatius, His Conclave*, the English poet and intellectual John Donne (1572–1631) imagines the Jesuit father (a figure he hated with a virulence) attempting to convince the devil to reject the teachings of Copernicus (as the Jesuits had done) on the grounds that Copernicus's teachings were all true. In this fantasy, written when Christopher Clavius was still at large in the Vatican, Ignatius meets up with the spirit of the long-dead Copernicus and discusses possible candidates for eternal damnation:

If therefore any man have honour or title to this place in this matter, it belongs wholly to our Clavius who opposed himself opportunely against you, and the truth, which at that time was creeping into every man's minde. Hee only can be called the Author of all contentions, and schoole-combats in this cause; and no greater profit can bee hoped for heerein, but that for such brabbles, more necessarie matters bee neglected. And yet not onely for this is our Clavius to be honoured, but for the great paines also which hee tooke in the Gregorian Calendar, by which both the peace of the Church, and Civil businesses have beene egregiously troubled: nor hath heaven it selfe escaped his violence, but hath ever since obeied his apointments: so that S. Stephen, John Baptist & all the rest, which have bin commanded to worke miracles at certain appointed dates . . . do not now attend till the day come, as they are accustomed, but are awaked ten daies sooner, and constrained by him to come downe from heaven to do that businesse.

Whether Clavius ended up in hell or heaven is not for us to know. Whatever he did or did not do, however enraging he was to the people of his time, we can at least offer him some posthumous thanks for the useful part he played in the history of calendar reform.

After all, the Clavius/Lilio team succeeded in addressing the two burning problems that calendar reformers had been unable to resolve for centuries. Firstly, there was the issue of how to adjust the regulation of the Julian calendar so that it would not err so much in the future, and, secondly, there was the immediate problem over what should be done about the 11-day slip that had built up since 45 BC and was currently contaminating the calendar. Their solution to the first problem was an obvious but drastic one – remove the 10 extraneous days – cut 'em out like a cancer. A papal decree was duly issued to eradicate 5 to 14 October 1582 from the calendar. 'Let it be done that after the 4th October, the following day shall be the 15th. Amen.' Imagine the chaos and civil disorder that must have followed such a proclamation:

'Excuse me. I have booked my ticket for the eleventh, but there was no eleventh, so can I travel today?'

'It deren't be t'eleventh today, modom.'

'Yes, I know there was no eleventh. What I am asking is if I can travel today instead.'

'It be twenty-first October today, modom. Ee should 'ave used that ther' tickey ont t'eleventh, modom.'

'But there was no eleventh, you fool, don't you understand? Today was going to be the eleventh, but they made it the twenty-first instead. Oh, please, for the love of God, may I travel today?'

'Yer can't be atravellin' on a tickey as were issued fer t'eleventh, modom, 'cos dates is dates, n'ther's an end to 'ee.'

Not every country jumped to attention as quickly as Pope Gregory and his agent, Clavius, had hoped. Some left a sensible pause to gauge the effect. Others, teeming with anti-Catholic virulence in an age of reformation, were not willing to show themselves to be servants to the Vatican. Why should they snap to attention at a papal bull – especially one as potentially destructive to the smooth engine of civilian affairs as this. In Frankfurt there were anti-papal riots in the streets, with Pope Gregory accused of stealing the days from the people. Particularly angry were the Protestants in countries like Germany who were outraged that the Pope should be able to control their lives from the Vatican. To them he was the 'the Beast', 'the Antichrist', whose bombastic ways were an affront to their own hard-earned sense of justice and fair play. One furious German professor wrote:

> We do not recognise this calendar-maker, just as we do not hear the shepherd of the flock of the Lord, but a howling wolf ... All his fatuous, contemptible mistakes, his sacrilegious and idol-worshipping practices, his vicious, perverse and impious dogmas that are condemned by the word of God ... These little by little he will once more insert into our churches.

This is the voice of a religious and partisan man, terrified that the Pope is planning a takeover bid of Protestant Germany. For others, even for committed Catholics, there were religious problems. People who wished to pray on saints' days to protect their health, enrich their crops or augment their profit were horrified to see those days disappear and fearful lest the saint in question should seek revenge for their omission. People with sore eyes, for instance, may have been intending to pray to the patron saint of eye afflictions, the fourth-century eccentric, St Triduana, on her saint's day, the eighth of October. It is said that when a handsome prince desired Triduana because of her beautiful eyes, she plucked them out and gave them to him. We cannot be surprised that she died a virgin. The sixth of October is reserved for the patron saint of soldiers, pilgrims and prisoners, St Faith – she was roasted on a brazen bed and beheaded; her symbol is a gridiron. But it was not just soldiers, pilgrims, prisoners and people with sticky eyes who were caught in the tangle of this calendrical chaos. Tax, interest on loans, deadlines brought forward, birthdays, weddings, ceremonies both religious and secular – all of these were affected. In short, the whole shooting match was thrown into disarray.

Englishman William Coxe wrote:

This innovation was strongly opposed even among the higher classes of society. Many landholders, tenants and merchants were apprehensive of difficulties, in regard to rents, leases, bills of exchange and debts, dependent on periods fixed by the old style. Greater difficulty was, however, found in appeasing the clamour of the people against the supposed profaneness of changing the saints' days in the calendar, and altering the times of all the immovable feasts.

Coxe was not writing in immediate response to Pope Gregory's reforms, for those were nearly 200 years earlier. He was writing about the introduction of Gregory's reforms into England, where, following the smooth passing of a bill through Parliament, it was agreed that Wednesday 2 September 1752 should not be followed by Thursday 3 but by Thursday 14 September – a removal of 11 days to align the British with most of the other calendars of Europe.

Each country took its own time to sign up. England had agreed to make the changes as early as 1583. Queen Elizabeth, advised by her friend the astrologer John Dee, persuaded the powerful men of her entourage (Francis Walsingham and William Cecil in particular) to accept the changes. The Archbishop of Canterbury, Edmund Grindal, fuelled with a deep mistrust of the papacy, was cautiously urging a policy of apathy towards the measure. By dilly-dallying, the archbishop was able to postpone the changes in England for another 170 years. So long, in fact, that the British obstinacy in this matter became the laughing stock of Europe. 'The English prefer their calendar to disagree with the Sun than to agree with the Pope,' joked Voltaire. '*Cur Anni errorum non corrigit*,' wrote another, '*Anglia notum, Cum faciant alii; cernere nemo potest*' ('Why England doth not the years known error mend, when all else do, no man can comprehend').

Other countries joined in dribs and drabs. England was careful not to repeat the errors of others. Flanders idiotically announced the change for 21 December 1582, which meant that all the Belgians missed their Christmas that year. In Bristol, England, 170 years on, violent riots broke out in which several people were killed and the song of the times, set to the tune of 'Old Tom Tiddly', went: 'In seventeen hundred and fifty-three/The style it changed to popery.' Most people in England at that time did not even know what 'popery' meant, but it stood for a concept they had been trained to loathe and to fear since their youth, and any excuse to rally against it was taken up eagerly.

Another form of protest was to carry on with normal life as though the reforms had never been introduced. This meant holding back on Christmas Day and eating the goose with all its stuffing and attendant trimmings on the following fifth of January, which would have been Christmas old style. One old man averred that the Glastonbury Thorn, a twig of *Crataegus*, or hawthorn, which traditionally flowered on Christmas Eve, 'contemptuously ignored the new style and burst into blossom on 5 January, thus indicating that Old Christmas Day should alone be observed, in spite of an irreligious legislature'. To the old believers, this hawthorn twig was none other than the sacred staff of Joseph of Arimathea planted there one day on his way to Glastonbury.

Other countries were even slower to accept the changes, but as greater parts of the world did join up so the need for the remaining countries to slacken their obstinacy and embrace the new system became ever more urgent. Protestant Germany fully accepted Pope Gregory's reforms in 1775; France, after her brief and fancy revolutionary digression, rejoined in 1806. Japan accepted the calendar in 1873, Russia in 1917 and China in 1949.

Removing the extra days was only one of Pope Gregory's concerns; the other was recalibrating the calendar so that the problem would never occur again – or at least not in a very long time. Clavius and Lilio promoted a beautifully simple solution. The great problem with Caesar's calendar was that it was too long. Looked at simply, it contained too many leap years. By inserting a leap day every fourth year, Caesar's year was 365.25 days long, so to reduce it down to the size of the true tropical year (the time it takes the earth to orbit the sun starting and ending with the March equinox) they needed to lop 0.0078 of a day off each year (0.78 days off a century or 3.12 days off every 400 years). History does not relate how much squabbling took place between Clavius and Antonio Lilio during the period of these great reforms, but Luigi's solution had an elegance that made it easy to understand and therefore (most important for Clavius) it would be easy for his Holiness to explain to his flock.

Leap years, it was decreed, should not apply to century years (i.e. 1700, 1800, 1900, etc.) unless the century year in question is exactly divisible by four (i.e. 1600, 2000, 2400, etc.). That put the calendar correct to within 25.9 seconds per year – not perfect, but a vast improvement. In every 2,800 years, if we are still using the system at all, which is unlikely, we shall have gained a day. By AD 4382 some busybody will no doubt pass an international decree ordering the removal of a single day from our calendars. If that does happen the computer chaos (far worse than the millennium bug) will bring the whole of civilisation crashing to its knees or, as Shakespeare put it:

> There is a tide in the affairs of men,
> Which, taken at the flood, leads on to fortune.

Antonio Lilio was rewarded for his own and for his deceased brother's work with an exclusive licence from the Pope to print and distribute the new calendar, including instructions on how to apply it. This was potentially a very lucrative deal. He could have built himself a palazzo with oils by Michelangelo and busts by Bernini to brighten the *piano nobile*; he could have bought himself lands and a title. His offspring might have married into the Medicis or the Farneses, enabling the lustrous name of Lilio to shine like a beacon for generations to come. But it was not to be. Lilio failed to print the first lot of calendars on time, and when they did arrive there weren't enough of them to go around. Furious at the added disarray that Lilio's incompetence had caused, the Pope, prompted by the unctuous Clavius, confiscated Lilio's printing licence and nothing was heard from the fellow again.

The Gregorian calendar does not entirely wrap up our story, even though it brings us to the system currently in use. The leap-year system may provide an effective method of keeping the calendar year synchronised with the solar year, but is it the best solution to the problem? People who care about these things say that it is not. February is a mess, with its scheme of 28 days or 29 days in a leap year, to which complaint may be added that the numbering of the days in the months appears to be almost random and very hard to remember without recourse to the childish mnemonic: 'Thirty days hath September, April, June and November. All the rest have 31, excepting February alone, which has but 28 days clear and 29 in each leap year.' There are many different versions to that, of course, some of which scan moderately better. Even so, it is still a rotten piece of doggerel.

Another problem with the Gregorian calendar which we are presently using is that the days of the week (Monday, Tuesday, Wednesday . . .) seem to fall on different days of the month in a permanently out-of-kilter cycle which repeats itself only once every 28 years. Would it not be more convenient if the first of January always fell on the same day of the week and that the other days of the year followed suit? Of course it would, but the problem remains: even if there is a better system, how on earth do you get the squabbling, fractious world to agree to put it in place? All the major calendar reforms of the past have been put in place by emperors, popes or potentates. All they had to do was to click their fingers and, barring a few skirmishes and dissenting protestations here and there, the thing was done. Nowadays dictators operate only in small countries, which, where calendar reform is concerned, is not a very good thing. Thousands of ideas have been designed to make the calendar better, but who is nowadays powerful enough to make the choice and push a reform through? The President of the the United States of America is nominally the most powerful person in the world, but he cannot get a world calendar reform past his legislature in the way that Julius Caesar or Pope Gregory could.

The neatest and most effective suggestion for calendar reform since the sixteenth century was invented by a Roman Catholic priest called Abbé Marco Mastrofini in 1834. He was born in the Roman district of Montecompatri in 1763 and died in Rome in 1845. He was professor of mathematics and philosophy at the college at Frascati and wrote many books on mainly ecclesiastical matters. He is best known for his treatise on usury. Pope Clement v at the Council of Vienne in Dauphiné in 1312 had announced that 'any one who shall pertinaciously presume to affirm that the taking of interest for money is not a sin, we decree him to be a heretic fit for punishment'. It was at the same council that he suppressed the order of the Knights Templar. In the nineteenth century Abbé Mastrofini tried to argue a case for Christians to lend money for interest. He was careful not to be too 'pertinacious' about it. His idea concerning the calendar, now called the World Calendar, has elicited an enormous ground swell of support since he first proposed it in 1834. Many books have been written about it, and religious leaders, presidents and the United Nations have all convened to discuss it. With almost universal approval for the World Calendar, why is it that these simple undisruptive changes have not been made? Time and again they have been blocked by zealous religionists who have prevailed upon the authority of the state to enforce their convictions. As Elizabeth Achelis, one of the driving forces behind the promotion and one-time president of the World Calendar Association, explains:

Opposition to calendar reform in a proposition that is astronomical and mathematical comes from the interference of certain religious groups. They fail to understand the universal and scientific character of the calendar and that it belongs to all nations, peoples and races. It does not deal with religious belief, dogma, theology, tradition, myth or orthodoxy. Therefore a tradition formulated in the time of the Babylonian captivity should not be made a reason for unreasonable adherence by Orthodox Jewry to their theory of the unbroken continuity of the seven-day week.

The system of the World Calendar is blissfully simple and easy to follow. The year is divided into four equal quarters of three months each. Each quarter contains 91 days; the quarters are made up of one month of 31 days and two months of 30 days each (see Table 2 overleaf). The months retain their original names. The first month of each quarter begins on a Sunday; the second month starts on a Wednesday and the third month of each quarter begins on a Friday. This keeps the same day of the week to the same date every year. (Bad luck if your birthday falls on Friday the thirteenth!)

This adds up to 364 days. To complete the scheme, a 'World's Day' holiday is intercalated every year between the last day of the old and the first day of the new year, i.e. between the thirtieth of December and the first of January. That makes 365, and then the leap-year day, another 'World's Day', is added once every four years after the thirtieth of June.

There are a great many reasons why this is better than the current system. The last big push for the World Calendar ended in failure. The US government sent a very pompous and sinister letter to the United Nations in March 1955, giving an indication of how impossible it might ever be, in this age-frightened democracy, to improve the outdated system that we have:

The United States Government does not favour any action by the United Nations to revise the present calendar. This Government cannot in any way promote a change of this nature, which would intimately affect every inhabitant of this country, unless such a reform were favoured by a substantial majority of the citizens of the United States acting through their representatives in the Congress of the United States. There is no evidence of such support in the United States for calendar reform. Large numbers of United States citizens oppose the plan for calendar reform which is now before the Economic and Social Council. Their opposition is based on religious grounds, since the

Table 2:

The World Calendar

JANUARY	FEBRUARY	MARCH
S M T W T F S	S M T W T F S	S M T W T F S
1 2 3 4 5 6 7	1 2 3 4	1 2
8 9 10 11 12 13 14	5 6 7 8 9 10 11	3 4 5 6 7 8 9
15 16 17 18 19 20 21	12 13 14 15 16 17 18	10 11 12 13 14 15 16
22 23 24 25 26 27 28	19 20 21 22 23 24 25	17 18 19 20 21 22 23
29 30 31	26 27 28 29 30	24 25 26 27 28 29 30

APRIL	MAY	JUNE
S M T W T F S	S M T W T F S	S M T W T F S
1 2 3 4 5 6 7	1 2 3 4	1 2
8 9 10 11 12 13 14	5 6 7 8 9 10 11	3 4 5 6 7 8 9
15 16 17 18 19 20 21	12 13 14 15 16 17 18	10 11 12 13 14 15 16
22 23 24 25 26 27 28	19 20 21 22 23 24 25	17 18 19 20 21 22 23
29 30 31	26 27 28 29 30	24 25 26 27 28 29 30 World's Day

JULY	AUGUST	SEPTEMBER
S M T W T F S	S M T W T F S	S M T W T F S
1 2 3 4 5 6 7	1 2 3 4	1 2
8 9 10 11 12 13 14	5 6 7 8 9 10 11	3 4 5 6 7 8 9
15 16 17 18 19 20 21	12 13 14 15 16 17 18	10 11 12 13 14 15 16
22 23 24 25 26 27 28	19 20 21 22 23 24 25	17 18 19 20 21 22 23
29 30 31	26 27 28 29 30	24 25 26 27 28 29 30

OCTOBER	NOVEMBER	DECEMBER
S M T W T F S	S M T W T F S	S M T W T F S
1 2 3 4 5 6 7	1 2 3 4	1 2
8 9 10 11 12 13 14	5 6 7 8 9 10 11	3 4 5 6 7 8 9
15 16 17 18 19 20 21	12 13 14 15 16 17 18	10 11 12 13 14 15 16
22 23 24 25 26 27 28	19 20 21 22 23 24 25	17 18 19 20 21 22 23
29 30 31	26 27 28 29 30	24 25 26 27 28 29 30 World's Day

introduction of a 'blank day' at the end of each year would disrupt the seven-day sabbatical cycle.

Moreover, this Government holds that it would be inappropriate for the United Nations, which represents many different religious and social beliefs throughout the world, to sponsor any revision of the existing calendar that would conflict with the principles of important religious faiths.

This Government, furthermore, recommends that no further study of the subject should be undertaken. Such a study would require the use of manpower and funds which could be more usefully devoted to more vital and urgent tasks. In view of the current studies of the problem being made individually by governments in the course of preparing their views for the Secretary-General, as well as the previous study by the Secretary-General in 1947, it is felt that any additional study of the subject at this time would serve no useful purpose.

Resumed Nineteenth Session, Agenda item 21
E/2701, 22 March 1955.

The World Calendar, the Julian calendar, the Gregorian calendar, Clavius, Caesar and all the calendar-makers based their calculations on the measurement of the tropical year. But there is another measurement for the year which is different from the tropical year, which is reckoned not by the sun crossing the equator but by the sun returning to exactly the same position in relation to the fixed stars. This is called a 'sidereal year' or a 'star year'. The term 'fixed stars' is a little bit misleading here. The idea of 'fixed stars' dates from ancient times when people believed that the stars did not move around in the universe, unlike the planets, which were known as the 'wandering stars'. In the early eighteenth century, however, the motions of the stars were measured.

We can take sidereal measurements for the month and for the day, as well as the year. In each case we will discover that sidereal time differs slightly from the length of days, months and years as they are normally measured in relation to the earth. This is not difficult to understand. In fact, you can do it yourself by following these simple instructions:

1. Get two huge megaliths. You will need one tall one and one shorter one. If you cannot find them close to hand, plenty are still available in the Prescelly Mountains of west Wales. Choose your megaliths with care and roll them on cut wooden logs over hill and dale to your home.

2. Next morning: get a stopwatch.

3. Stand the tallest megalith in the middle of your lawn and place yourself on the other side of it from the rising sun. Then place the shorter megalith in between you and the tall megalith (you will have to be quick, especially if you are living on the equator). Now practise moving yourself to a sightline where the top of the small megalith is exactly aligned to the top of the tall megalith. (You may have to squat a bit.)

4. Keep squatting until the last fraction of the circumference of the sun rises above the alignment of the tops of your two stones. At that exact moment press GO on your stopwatch. (Be careful to wear very dark glasses or you might blind yourself.)

5. Leave the stones exactly as they are overnight and allow the watch to keep running.

6. Next morning, without moving the stones, align yourself, in the same way as before, to the rising sun. When the last fraction of the circumference of the sun rises above the alignment of the tops of your two stones, press STOP on your stopwatch. Well done. You have just measured your first solar day. You should have a reading of something close to 24 hours, depending on where you are in the world and what time of year it is. Now, if you repeat that process every day for 365 days, take a reading for each day, and then divide by 365, you should come up with a figure for the mean solar day. If you have done it properly, your calculator

Three different views of the Creation: (*top*) From Mexican painter Diego Rivera (1886-1957); (*bottom left*) From an ancient Egyptian tomb painting; (*bottom right*) The Hindu egg creation myth.

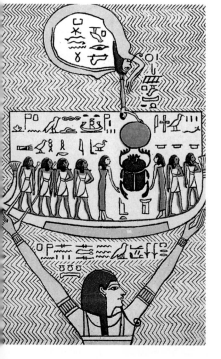

Zeno of Elea, ancient Greek philosopher, whose paradoxes suggested that a runner can never reach the finishing post.

An ancient Greek athlete trying to defy Zeno's prediction.

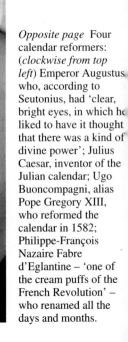

Opposite page Four calendar reformers: (*clockwise from top left*) Emperor Augustus who, according to Seutonius, had 'clear, bright eyes, in which he liked to have it thought that there was a kind of divine power'; Julius Caesar, inventor of the Julian calendar; Ugo Buoncompagni, alias Pope Gregory XIII, who reformed the calendar in 1582; Philippe-François Nazaire Fabre d'Eglantine – 'one of the cream puffs of the French Revolution' – who renamed all the days and months.

March: A scene from the most lavish Book of Hours, the Duc de Berry's *Très Riche Heures* (early fifteenth century).

Giovanni Antonio Pordenone's allegory of time (early sixteenth century).

A Dance to the Music of Time by Nicolas Poussin (1594–1665).

King Herod's most brutal crime: *The Massacre of the Innocents* by Sebastiano Mazzoni (1611-1678).

A group of learned astronomers in seventeenth century France, from Gustave Doré's *La Fontaine's Fables*.

An early nineteenth century fantasy depicting a pagan festival at Stonehenge.

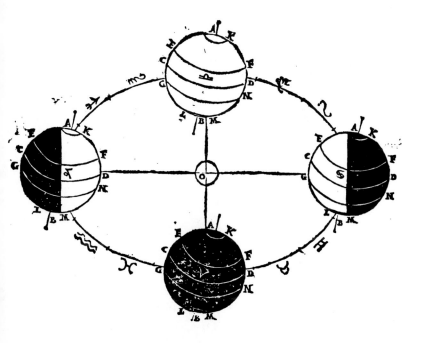

Copernicus's heliocentric schema, as illustrated in a book by Galileo of 1632.

Above Albert Einstein, whose theories of relativity exposed time as the fourth dimension of our physical universe.

Above right Hermann Minkowski, the scientist who developed the theory of spacetime.

Right An apocalyptic message stands out in a crowd.

should show 24 hours, three minutes and 56.555 seconds. If it does not, try again.

To measure the sidereal day you need to go out on two consecutive cloudless nights and arrange a convenient sightline between yourself, the tops of your megaliths and a prominent fixed star.

By measuring how long it takes for a fixed star to return to exactly the same position (relative to the aligned tops of your megaliths) on two consecutive nights, you can get a reading for a sidereal day, which you will probably notice is shorter than the solar day by about four minutes. The reason for this is that the earth, as it rotates on its axis, is also orbiting the sun; so if we measure the rotation of the earth by watching the sun, when the earth has turned around once the sun will appear to have moved eastwards, and so the actual moment of noon (or the point where your megalithic alignment has returned to the same position as on the previous day) will in fact occur four minutes after the earth has completed its 360-degree rotation.

Sidereal months can be measured by noticing a distant star in its relation to the moon. For instance, when a star appears to be butting up against the edge of the moon, the same position will not be repeated for exactly one sidereal month and, once again, this will differ from the lunar month for the same reasons of rotation and orbit that affect the solar day.

The solar and sidereal years are harder to measure, even though the principle remains the same, because of the need to take accurate readings for a whole year. This, the great Ptolemy and the Arab astronomer al-Battani failed to achieve in their own time. Copernicus, working like a hermit high up in a turret set in the fourteenth-century defensive city walls of Frauenberg, noticed discrepancies between his own readings and those of Ptolemy and al-Battani hundreds of years earlier. Not realising that they were in the wrong, Copernicus incorrectly assumed that the earth must move in irregular motions. This small mistake, however, did not prevent him from measuring the year far more accurately than it had ever been measured before. His tropical year came in at 365 days, five hours, 49 minutes and 29 seconds – only 43 seconds out.

Copernicus was born in eastern Poland in 1473. His uncle, Bishop of Ermeland, wanted him to enter the Church. After studying at Cracow University and later in Padua, he started to gain a reputation as a proficient and informative stargazer. So much so that by 1514 Pope Leo X had invited him to Rome to give his suggestions on how to reform the calendar. Unfortunately, Copernicus had nothing interesting to say on the matter.

From 1512 Copernicus lived in Frauenberg, near Danzig, where he made his greatest discovery, proving, to his own satisfaction, that the earth orbited the sun and rotated on its own axis. This heliocentric view was not entirely original. The theory is attributed in the first instance to Aristarchus of Samos, who lived between 310 BC and 231 BC. This ancient astronomer might have the honoured place in the history of science that Copernicus presently holds if only his book, in which he is purported to have proposed a Copernican solar system (where the planets, including the earth, revolve around the sun and the earth turns on its axis once every 24 hours), were not lost. To Aristarchus's detriment, the only surviving work of his is *On the Sizes and Distances of the Sun and the Moon*, in which he calculates the diameter of the earth to be a mere 180 miles (out by a factor of 65!) and proposes the conventional and inaccurate view that the earth, not the sun, is at the centre of the system. It is thought that he may have come to the Copernican theory only after writing this book. In any case, he is known to have got there first. Archimedes (a contemporary of Aristarchus), in a letter to Gelon, the hugely powerful military king of Syracuse, mentions that Aristarchus has written 'a book containing certain hypotheses, namely that the fixed stars and the sun do not move, that the earth revolves about the sun in the circumference of a circle, with the sun lying in the middle of the orbit'. The great book in which Copernicus published his heliocentric theory, *De Revolutionibus Orbium Coelestium*, was not presented to the world until after his death. In fact, it was not even presented to the author until the moment of his death, by which stage the astronomer's mind had more or less gone. It has been said that the last thing he set eyes on before breathing his last breath was the galley-proof of his new book. But in some ways Copernicus's death was fortuitous, for his heliocentric view was heretical and the punishment for heresy was usually several days of torture followed by ignominious public execution, for, according to Jewish and Christian teaching, the world was at the centre of the planetary system and was ordained to be that way in the scriptures.

These clerics refer, to assert their authority, to a puzzling episode in the Bible in which Joshua, Moses' successor, orders the sun to stand still while he slogs it out on the battlefield with some Amorites:

> Then spake Joshua to the Lord in the day when the Lord delivered up the Amorites before the children of Israel, and he said in the sight of Israel, 'Sun, stand thou still upon Gibeon; and thou, Moon in the valley of Ajalon.' And the sun stood still and the moon stayed until the people had avenged themselves upon their enemies. So the sun stood still in the midst of heaven, and hasted not to go down about a whole day. And there was no day like that, before or after it.

Copernicus was now telling the world that there *was* a day like that both before and after. In fact, every day that has ever been was, is and will be like that. Does this mean that Joshua is a fool, that the scriptures have got it wrong? Impossible! The Church knew that the scriptures were always right. If the Bible says that Joshua ordered the sun to stay still and that the burning planet, for the first and last time in its life, did indeed stay still, then the Bible must be right and Copernicus, with all his fancy scientific tools, must be a heretical antichrist.

The Church of Rome was particularly keen, in the years after Copernicus's death, to suppress his work, but it was not just in the Christian religion that the geocentric view of the solar system was upheld as a sacred tenet. A creep called Cleanthes went around whipping up the mob against Aristarchus in the fourth century BC. According to Plutarch, Cleanthes

> thought it was the duty of the Greeks to indict Aristarchus of Samos on the charge of impiety for putting in motion the Hearth of the Universe [he means the earth], this being the effect of his attempt to save the phenomena by supposing the heaven to remain at rest and the earth to revolve in an oblique circle, while it rotates, at the same time, about its own axis.

Whether or not Copernicus knew of Aristarchus's heliocentric theories is unknown, though he may well have come to them through Seleucus, the first and the last of Aristarchus's followers to promote his view. At any rate, it is known that Copernicus confirmed the view, using his own observations and calculations. This much is evidenced by the scribbles he left behind. Interestingly, *De Revolutionibus Orbium Coelestium* turned out not to be the bombshell one would have expected when it was published in 1543. In the first place, most people could not understand it, and editions were issued with a patronising preface claiming that Copernicus's views, herein published, were all conjectural. It was not until Galileo Galilei (1564–1642) restated the same heliocentric theories as Copernicus that the Church authorities really woke up to what was going on. Galileo, who had advanced the telescope as a key instrument in astronomy, was denounced as a heretic.

To Copernicus's heliocentric model many enlightened scientists were able to add their own discoveries. Most important of these was the stinking flea-bitten astronomer Johannes Kepler (1571–1630), son of an oafish mercenary soldier and his cantankerous wife (a publican's daughter who was later tried as a witch). Kepler was supported at one time by the bad-tempered Danish astronomer Tycho Brahe (1546–1601). Brahe was a wealthy aesthete, obsessed from the age of 14 with watching the stars and surrounding himself with attractive objects. With royal money he built himself an island observatory called Uraniborg

(Castle of the Heavens) from where he catalogued the positions of 777 stars. He married beneath his station, which caused a rumpus at the time, and was able to put up with Kepler's awful smells only because he could not notice them: his nose was knocked off in a fight as a teenager and he had constructed himself a false one (which did not work) made of silver. Together Kepler and Brahe worked on joint astronomical observations. Kepler's most important discovery was the elliptical movement of the planets. Up until then it was thought that they moved in perfect circles. Later it was shown that gravity was responsible for their elliptical trajectories. In the meantime Kepler's proof helped to explain small irregularities in the observation of the equinoxes.

An equinox, you may remember, is the time or date at which the sun crosses the celestial Equator and may be identified as the two days of the year in which both day and night are of equal length. It is the solar opposite, if you like, of a solstice. A solstice is so called because at these times the sun reaches its extreme northern and southern points in the ecliptic and appears to stand still; the word comes from the Latin *sol* ('sun') plus *sistit* ('stands') and is identified as either the longest or the shortest day of the year. The ecliptic, by the way, is the technical term for the plane of the sun's orbit. The two equinoxes occur each year in September and March, the two solstices each year during December and June.

The year, therefore, can be measured between either of the two equinoxes or either of the two solstices. The problem we have, or, to be accurate about this, the problem very pedantic scientists have, is to identify one lasting definition for the year which is not subject to variation. This was nearly achieved in the World Conference of 1967, but the sort of information it yielded can be of interest only to scientists and their laboratory computers. We are talking here about fractional figures that are either so infinitesimally small or so perilously large that the human brain cannot possibly comprehend their meaning except by comparison (i.e. are they greater or are they smaller?) with other numbers.

The official time-measurement for the year is so obtuse that it pains one to fulfil the need to purvey it here. We touched on atomic time in the discussion of seconds – *'Un fastidio m'ingrombra la mente'* ('A terrible bother clouds the mind') as the great Italian poet Giacomo Leopardi would have put it. For the record, however, expert David Ewing Duncan explains the meat of this issue with admirable concision in his book *The Calendar*:

In 1967 the rate of cesium's pulse was calibrated to 9,192,631,770 oscillations per second. This is now the official measurement of world time, replacing the old standard based on the earth's rotation and orbit, which had used as its base number a second equal to $1/31,556,925.9747$ of a year. This means that under this new regime of cesium, the year is no longer officially measured as 365.242199 days, but as 290,091,200,500,000,000 oscillations of Cs, give or take an oscillation or two.

So there you have it!

CHAPTER IX

DECADES
Decades

Pythagoras, the ancient Greek philosopher and mathematician, is now famous above all for the brilliant classroom theorem of the right-angled triangle. It states that the square of the hypotenuse (the longest side) is the sum of the squares on the other two sides. If we only ever retained one thing from our school maths classes, Pythagoras's theorem must surely be it. It is possible, however, that Pythagoras did not actually invent the theorem himself, but that is a small matter. He was a figure of towering importance, not only to the people of his time, but also to scholars of later ages. Pythagoras has remained a key figure in the history of Western mathematics and philosophy for 2,500 years.

In 510 BC, Pythagoras settled in the city of Croton, famous then for its medicine, where his teaching soon gained a following of earnest disciples. But the regular citizens of Croton soon tired of him, finding him and his sect scary and eccentric, so Pythagoras and his flock all moved out to Metapontion. At Metapontion, his fame grew to such proportions that many believed him to have magic powers, and he was frequently visited by people with incurable ailments seeking a miracle cure. He worked hard at his mathematics but, as Aristotle later put it, he 'condescended' to instigate his own religion – an ignoble thing to do. In fairness to Aristotle, it was the strangest of all religions which advocated, among other things, reincarnation and abstinence from beans.

In Shakespeare's *Twelfth Night*, Feste the clown asks: 'What is the opinion of Pythagoras concerning wild fowl?'

'That the soul of our grandam might haply inhabit a bird,' replies Malvolio.

That was the tip of the iceberg as far as the fuller face of Pythagorean philosophy was concerned. To be a true follower of Pythagoras you were supposed to obey a set of rules that the great man himself had decreed in stone: 'Never eat beans' is the most famous stricture of the Pythagorean cult, but the other, lesser-known tenets of his faith are even more bizarre. Never allow swallows to perch on your roof, for instance; never pick up anything that has fallen down; never step over a crossbar or sit on a quart measure; do not break bread, or eat from a whole loaf; and, perhaps the strangest of all, never touch a white cockerel.

To these, Pythagoras added other fervent beliefs concerning numbers. One was good, the purest of all numbers representing unity; two stood for diversity and was therefore chaotic, striven and evil – a superstition taken to heart by the later Romans who extended it to include all even numbers. Curiously, Pythagoras held four to be a perfect number as it was the first square (two times two). Three, being the union of unity and diversity (one plus two) was believed to be the number of perfect harmony; five represented nature and art, six stood for justice. Seven was the medical number when supposedly everything went wrong. Nine was the same, as the later Roman historian Levinius Lemnius recalled: 'There are two years, the seventh and the ninth, that commonly bring great changes in a man's life, and great dangers; wherefore 63, that contains both these numbers multiplied together, comes not without immense dangers.'

These and many other curiously strict rules were followed for several hundred years by Pythagoras's loyal disciples. After that many of them lost heart, and their desire to tidy things up which had fallen to the floor, or to stuff their faces with delicious oiled beans got the better of them. One by one, they deserted the sect.

Nichomachus of Gerasa was an exception; he stuck with it to the end of his life, expounding Pythagorean principles in his seminal *Introduction to Arithmetic* – a work which was later to prove one of the most influential books on mathematics of the Middle Ages.

Nichomachus had, it is believed, taken from Pythagoras's teaching the view that 10 (like four) was a perfect number. It has generally been assumed that we count in base 10 for one specific reason only, which dates back to the times when humans used their 10 digits to count and to barter in the pursuit of market trade. But Nichomachus had it the other way around. Pythagoras had taught him that 10 was 'the first-born of the numbers, the mother of them all, the one that never wavers and gives the key to all things'. He fervently believed that God liked the number 10 so much that he had ordered the Creation around it, and that the Good Lord gave man 10 fingers and 10 toes so that he might be able to use this sublime number in his counting and arithmetic.

Nichomachus and his ilk were deceived. If mankind had evolved with six fingers on each hand, Nichomachus would undoubtedly have fallen to his knees at the mere mention of the number 12. Likewise, if there had been just four fingers on his whole body, would he and all his contemporaries not be summing and subtracting under the 'divine' influence of base four? The decimal system is as old as civilisation. The Sumerians who used base 60 and the Egyptians, who favoured base 12, were odd ones out.

So decades, centuries and millennia are all purely human constructs. They have no connection with the phases of the moon, the orbits of the sun or the rotations of the earth, but are pure human concoctions, borne from our irrepressible love of the number 10, and our consequent perception of its usefulness. So the 'naughty nineties', the 'roaring twenties' and the 'swinging sixties' would never have existed had it not been for the fact of our having four fingers and a thumb on each hand. This might seem strange at first, but we must remember that the superiority of the human race over other animal species owes as much to the dexterity of our fingers as it does to the sophisticated workings of our brain. We owe much more to the finger than we can imagine.

In the early sixteenth century, Leonardo da Vinci, painter, sculptor, architect and engineer *extraordinaire*, measured in fingers' breadths (about 2cm): there were four of these to a 'palm' and six palms to a 'cubit'. In England in the Middle Ages, the word 'finger' was also used to delineate a measurement of exactly an eighth of a yard (11.4 cm). The idea probably derived from an ancient Roman unit of length known as the *digitus*. The English also measured units known as 'nails', being half of a finger; 'hands' (still used today for measuring the height of horses), and 'feet' (12 inches). The use of all of these has now, ironically, been superseded by the ultimately 'digital' decimal system.

And so to decades. The word 'decade' is itself a contraction of the original phrase, 'decade of years'. We hardly use the word to mean anything else any more, but traditionally a 'decade' or 'decad' were adjectives used regularly to refer to anything from people to pumpkins and fieldmice. There are precious few recorded incidents of the terms forties, fifties, sixties, etc., being used before the twentieth century. Until the end of the nineteenth century the familiar way to divide the centuries was not by decades but by referring to the reigns of kings and queens. When the term 'decade of years' was used originally (the first known written example comes from T. Hutton's *Reasons for Refusal* of 1605), it never referred to the specific phenomenon of decades as we now know them (i.e. divisions of the century) but to any group of 10 years. Thus in Rawlinson's *Ancient History*, 'The war might still have continued for another decad of years.' The word was more commonly spelt without the silent 'e' up until the twentieth century.

It must be remembered that the general populace had little idea of dates and times at least until the eighteenth century, and the idea of decades and centuries as we presently view them had not entered the broad public consciousness until the 1890s. We have seen how Pope Gregory's calendar reforms of 1582 caused great dismay throughout Europe, but that is not to suggest that all people were so aware of the date in 1582 that date-sensitivity was the sole cause of their wrath. Human nature is never that simple. At that time, anti-papal sentiment was far stronger in Germany and England than date-love was. Luther and the Protestants were convinced that the Pope was the Antichrist mentioned in the Book of Revelation, the Beast of number 666. His changing the dates was therefore an act of devilry over which they could all riot and rave, but few of them really understood what the 1582 reforms were about. If people were not aware of dates, it was because they did not on the whole need to use them. In rural communities it was sufficient to know when Christmas and Easter would fall, but the Church would tell them that, for the Church was the traditional custodian of the village calendar. For the ordinary people of 1582 the calendar was a prohibitively complicated mechanism which involved a degree of knowledge in mathematics, astronomy and ancient history, not to mention a deep understanding of the scriptures, which was all way beyond them. With the Church in charge, maybe the people felt safe. God was ultimately providing a calendar which worked, and on Sunday the local priest was sure to inform them if anything important like Christmas or Easter were lurking round the corner. But in God's supposedly safe pair of hands, the calendar was prone to vacillation. Even after Pope Gregory had implemented his reforms, the average lay citizen had no idea what was going on. Take one example:

In England the Gregorian calendar was officially introduced, as we have seen, in 1752. From the Middle Ages, dates were indicated by regnal years followed by a number which signified a date either before (or after) a given festival of the Church. So, for instance, the date we now know as 19 April 1514 would, in England, have been written down as the 'fourth day after Easter in the sixth year in the reign of King Henry VIII'. Obviously, a very cumbersome way of telling the date. Not only, in this case, is one required to remember the date on which Henry

VIII ascended the throne of England, but it also requires a computer-like memory to recall on which day Easter fell in that year. Other days of the year would be reckoned from St Crispin's Day (the twenty-fifth of October), the Gunpowder Plot (the fifth of November), St Philip and St James the Less (the first of May) and so on and so forth. Not only did regnal years change, but so did the feasts from which the days in any given year were supposed to be measured and, on top of all this, the beginning of the year itself was also moved about. No wonder the people could not be bothered with dates. 'We'll leave it to the Church,' they thought.

Not only are the regnal years and church festivals horrendously confusing but dating of English history has also been severely complicated by another problem: the moving around of the first day of the year. Any document more than 300 years old which has New Year's Day on it should be treated with extreme caution. It could have been written on the first of January, the twenty-fifth of March or Christmas Day, the twenty-fifth of December. Copies of the *Book of Common Prayer* published during the Reformation stated that:

> the supputation [reckoning] of the year of our Lord in the Church of England beginneth the Five and twentieth day of March, the same day supposed to be the first day upon which the world was created and the day when Christ was conceived in the womb of the Virgin Mary.

The twenty-fifth of March remained the first day of the year in subsequent editions of the *Book of Common Prayer* but the religious reasoning of the earlier editions was removed. The year in which the Gregorian reforms were introduced into England, 1752, was also the year in which the first of January was officially pronounced to be the first day of the year. To confuse historians even further, some people had already started using the first of January as New Year's Day before the bill was passed, so that dates referring to January, February and most of March pre-1752 might indicate either one of two years. So, for example, a document that reads 4 January 1683 will be transcribed by the high-minded historian as 4 January 1683/4.

It took over 100 years for the reforms of 1752 to sink in properly. By the end of the nineteenth century Queen Victoria's reign had been dragging on for more than 50 years and, with international trade and communications booming as never before, dating from royal coronations became obsolete in all but the highest affairs of state. By the 1890s people were able to cast their eyes backwards and flesh out their recent history in terms of Dionysian decades. The 1840s were nicknamed 'the hungry forties', referring to the period between 1844 and 1848 when most of Europe suffered a series of crop failures, most notably in Ireland where the staple food – the potato – was blighted and many thousands starved or were forced to emigrate. Sir Robert Peel, the British prime minister, tried to help by repealing the Corn Laws in 1846, which could maintain prices at an artificially high level. However, as supplies of corn remained short, the prices remained high and the poor struggled to afford basic staple foods. This combination of events was to lead to a series of revolutions throughout Europe in 1848.

The 1890s themselves came, in subsequent decades of the twentieth century, to be known as the 'naughty nineties'. In most of Europe and America these years marked the beginning of a new, freer era, which was to throw off the yolk and the tiresome constraints of the worst elements of Victorian puritanism. The 1890s were, above all, a decade for the ostrich-feather hat, for Aubrey Beardsley and his lewd artwork, the telegraph and the telephone, and, most exciting of all, the moving image. When Queen Victoria saw a film for the first time at the 1896 Lord Mayor's Show, she remarked in her rich Hanoverian accent: 'very tiring to the eyes but worth a headache to have seen such a marvel'.

In the twentieth century, particularly after the Second World War, the idea of named decades with characters of their own became something of a running obsession. It is interesting to observe, though, that the first two decades of the twentieth century, which cannot be easily named (unlike the 1920s, 1930s, 1940s, etc.) are the two decades of the century with the least easily identifiable characters. The years 1910–19, for instance, have never been known as the 'teens' nor have they ever been given a descriptive sobriquet like naughty, roaring or swinging. Terrible teens might have been appropriate for this was, of course, the decade of the Great War. The decade before that, 1900–1909, has no chance of a nickname for there is no word in the English language (like twenties, thirties or forties) to describe the numbers one to nine. The conclusion must be drawn that numerology (our excessive interest in the number 10, for instance) and language (our ability to classify in one word the years 1920–29 but not the years 1900–1909) are seemingly irrelevant factors which actually succeed in shaping and prejudicing our view of history and the way in which we interpret it.

There is another factor involved which can be added to numerology and language as guilty twister of history – namely music. When we cast our minds back over the twentieth century, for example, how do we instantly view each decade? Take the 1960s; they were the 'swinging sixties', the age of sexual liberation and the miniskirt and high-rise flats. Most of these are visual images, but with them surely comes the sound of the Beatles, the Rolling Stones and the Monkees. Likewise the 1950s are represented above all by Buddy Holly and the then trim Elvis; the 1920s conjures up the sound of foxtrot, charleston and Dixieland jazz – but why should this be? After all, there were millions of people who lived through the 1920s, the 1950s and the 1960s without ever bothering to listen to any of them. Why should rock'n'roll be the defining sound of the 1950s when most people who lived through that decade did not even like it? As Frank Sinatra said in 1957: 'Rock'n'roll is the most brutal, ugly, vicious form of expression . . . sly, lewd – in plain fact dirty.' The Spanish cellist Pablo Casals described it as 'poison put to sound – a brutalisation of both life and art'. But Sinatra and Casals were both expressing a preference; each had a passion for gentler, more subtle forms of music; forms which, in their view, were under threat of extinction from the noisy and unsophisticated exuberance of the new rock'n'roll. Others were more puritanical. The Revd Albert Carter is best remembered for his hysterical assessment of 1956:

> The effect of rock 'n' roll on young people is to turn them into devil-worshippers; to stimulate self-expression through sex; to provoke lawlessness, impair nervous stability, and destroy the sanctity of marriage.

In an even more extraordinary attack, American councillor, Asa Carter, from Alabama, said in the late 1950s something that would have undoubtedly cost him his job in the more sensitive climates of the ensuing decades:

> Rock 'n' roll is a means of pulling down the white man to the level of the 'Negro'. It is part of a plot to undermine the morals of the youth of our nation. It is sexualistic, unmoralistic, and the best way to bring people of both races together.

And yet if we think of the 1950s we do not have an instant image in our minds of puritanical people like the Councillor Carter and the Revd Albert Carter. That many of these priggish Carter people existed in the 1950s is beyond doubt, but it is the music which they despised and not their prim reaction to it that has characterised the decade. Why is this so? Why, for instance, when we think of the 1960s, is it the youth culture with its floppy hair and colourful dresses, jogging away to the strains of the Beatles that we imagine? What about the people who hated all that? The decade is known as the 'swinging sixties' not the 'censorious sixties'. Swinging in the 1960s was new; censorious behaviour, on the other hand, was as old as the hills. Even so, to describe the decade as 'swinging' shows a curious bias to music. In the twentieth century it looks as though music has played a unique role in presenting our history to us.

Take a look, for instance, at the 1920s. In a flash, this too is a decade that is quickly generalised. The charleston, a frenetic dance which involves kicking out your legs, turning in your toes, sticking out your heels and allowing your arms to twirl about in wild syncopation, took America by storm when it was featured in the black musical *Runnin' Wild* in 1923. By 1925 the craze had spread to London; within months it was all over Europe. The older generation predictably detested it. Middle-class mothers were particularly anxious for their well-brought-up young daughters to refrain from the dance. But the urge was too strong, as a charming song from the period makes plain:

> I wonder where my baby is tonight,
> I wonder why my baby doesn't write
> She left me 'cos I couldn't do the charleston, charleston
> She left me 'cos I couldn't do the charleston, charleston
> I am taking lessons now,
> I'll win her back somehow.
> I wonder where my baby is tonight.

The French, a nation always willing to take things one step further, had a charleston queen of their own. Josephine Baker was the 19-year-old lead dancer of *La Revue Nègre*, which later moved to the famous *Folies Bergère*. Baker's version of the charleston involved slapping her bare buttocks in time to 'Yes, sir, that's my baby!' followed by a topless mating dance with nothing more than a few stripy feathers to cover her other parts. Her dancing not only thrilled the lusty male audiences who came to see her but became something of a cause célèbre among the artists and intellectuals of her day. Picasso called her 'the Nefertiti of today'. (He was presumably reminded of a limestone bust dating from 1350 BC, now in a museum in Berlin, which does, it must be said, bear a striking resemblance to Josephine Baker.) Colette called her 'a most beautiful panther', while to Anita Loos she was little more than 'a witty rear end'.

And so the charleston and all that early jazz has come to define the 1920s as a decade. But what really happened in the 1920s? Was it all about speakeasies and foxtrots, cocktails, boaters,

Oxford bags and straight dresses? Of course not. The 1920s was a decade like any other. It had its tragedies and its triumphs, its energy and its ennui.

For the people living in the 1920s the decade cannot have been as it is now portrayed. Music is like a fire – it moves and therefore appears to be alive – but that is all illusion. When we read a history of the 1920s, when we see the pictures that were painted or the photographs that were taken, it is always the music more than anything else that appears to bring the era to life. What a strange irony that we can take a bogus division of time, based on our own love of the number 10 which is directly influenced by the number of digits on our hands, and then feed into this bogus period a falsely clarified picture of what it was all about. For those of us who were conscious of being about in the 1970s, the 1980s or the 1990s, did we not, at some stage, ask ourselves: 'How will this decade be seen in years to come? I cannot imagine. It is so disparate, there are so many competing styles, the music is probably not good enough to last, the fashions in clothes are too eclectic. What will it be that defines this decade in many decades to come?' And yet as the 1970s turned into the 1980s and the 1980s moved over to make way for the 1990s, so our view of these past decades became ever more channelled, more selective, clearer, yet more inaccurate. Most people alive now do not have personal recollections of the 1920s; those who were adult enough to form proper memories of those years have, in the main, had their memories clouded by things they have seen and read, not to mention the passing of time and the tricks that old age is known to play on the memory. And so all our memories are funnelled down these false paths. We will never know what the 1920s were really like and nor will anyone who lived through them ever experience those years again. Maybe there is something positive to be learned out of all this, a nugget of hope, perhaps? Is it in the nature of music, television or of human memory that gives us the images of the foxtrot and charleston before any other? What really happened in the 1920s?

The worst event of those years was surely the great earthquake in Tokyo and Yokohama which reduced the two cities to rubble, killing 300,000 people and turning more than a million others into homeless refugees overnight. This happened in September 1923. In Germany, the price of a loaf of bread soared from 250 marks in January 1923 to 201 billion marks by November of the same year. Hitler's Nationalist Nazi Party was picking up supporters in response to the collapse of the currency. In Britain, there was a general strike which brought the country to a standstill in 1926; in Vienna a state of emergency was called when riots broke out, started by Communist Party members who were suffering from high dudgeon because two men from the anti-socialist front had been acquitted of murdering two of their agitants; 69 children died in a cinema fire in Paisley in Scotland, and on 24 October 1929 the stock markets crashed in New York causing 11 people to commit suicide and leaving other investors all around the world miserable, shocked and destitute. These were the bleaker episodes.

On the brighter side the world land speed was broken twice, first by Malcolm Campbell in his 174.224-m.p.h. *Bluebird* in 1927, and then by Major Henry Segrave in his 231-m.p.h. *Golden Arrow* on Daytona Beach in 1929; Charlie Chaplin made *Goldrush*, Tutankhamun's tomb was uncovered by Howard Carter and Lord Carnarvon, while Mussolini, the Italian fascist leader, was shot in the nose by a female Irish aristocrat. It was a decade with its ups and downs, much like any other.

A thorough historian could easily show that the 'roaring twenties' were not particularly roaring. Lives had to be rebuilt after the Great War and while some (the better off) managed to express their relief in behaviour that may loosely be described as 'roaring', it was by no means universal nor did it necessarily span the whole decade. The same applies to the so-called 'swinging sixties'. That name was given in recognition of 'Beatlemania', which did not hit the United States until 1964. Similarly, the 'me decade' – a phrase coined by the American novelist Tom Wolfe to describe the ruthless ambition and pure selfishness of 'yuppies' (young urban professionals) in the 1980s – ignores the fact that the yuppy phenomenon was a theme particular to the later part of that decade when huge salaries were offered to very young brokers in Wall Street and the City of London.

As a unit of time, the decade holds a unique position. If, in theory, it existed from the moment the first calendar was invented, in practice it is a nineteenth-century idea used then, as now, almost exclusively to delineate units of time in the past tense. We frequently refer to minutes, years and millennia as units of either past, present or future time, but this is rarely the case with decades. In France, Italy and Germany, the same applies; the term refers to times that have already been and gone or, at best, we are presently living through. As a measuring tool, the decade is of moderate convenience, but the poetic sobriquets ('hungry', 'swinging', 'roaring', etc.) which time has attached to so many of them have created, for the general public, a distorted view of modern history and an artificial distance between our present and our recent past. But we are inveterate in our love for nostalgia and nothing will control that, so for us the decade has become the supreme unit of nostalgic time measurement. For 'the past', as the poet Edward Thomas so ruefully put it, 'is the only dead thing that smells sweet'.

CHAPTER X

SAECULA
Centuries

C enturies, even more than decades, are the symptoms of man's babyish excitement about numbers with noughts on the end of them. Neither the Sumerians nor the Egyptians would have been remotely interested in centuries, as we have already shown. For the former a period of 3,600 years, for the latter 144 years would have been far more absorbing than the awkward, lumpish hundred. We have already noted how in the Jewish faith the number seven takes special pride of place but there are very few numbers, except those that are prominent in the decimal system, over which we are prepared to fire our imaginations.

There is simply nothing special about centuries, except that they contain 100 years and to anyone using the decimal system 100 is a special, if not exactly a mystical number. As we look across the spectrum at all the time units which we currently use – minutes, seconds, hours – it becomes apparent that our whole system of time measurement is based upon specific numbers which at some moment in history, and for some reason, were held to be sacred, special, or important in some sort of way.

Perhaps the supreme example of our fallible human obsession with numbers can be found in the study of Gnosticism, a branch of mystic Judeo-Christianity which thrived on the Mediterranean coast and in northern Europe during the early part of the Middle Ages.

The Gnostics believed that if they could only find out the name of God, they could somehow enter on to his level, rise to a higher plane as it were. The name of God itself they conceded might never be known to them, but they lived in hope that one day they might at least discover the *formula* in which his name was contained. According to them, the formula was not exactly the same thing as the name itself; it was a code based on numbers, letters, symbols or a combination of all three that would contain the name and might, if they were lucky, lead them closer to the seat of almighty holiness.

The German poet, lately revealed to be a spy, Wolfgang von Goethe, would have suffered no truck from the Gnostics. 'Names are but noise and smoke,' he wrote, 'obscuring heavenly light.' Shakespeare's similar point is better expressed and consequently more often quoted: 'What's in a name? That which we call a rose by any other name would smell as sweet.' But nothing would put a tunnel-visioned Gnostic off his course. God had a name, and it had to be found. That was all there was to it. So how does the Gnostic set about finding the holy name?

First of all he needs to find a magic number. According to Basilides (a prominent Gnostic of the second century AD), 365 had magic powers, for it was the number of the minor gods who ruled over the days of the year and so 'Big God' – the main god, that is – must surely be the sum of all these smaller ones.

The next important task was to work out what name corresponds to the number 365. The Romans used letters for their numbering system (I is one, V is five, X is 10, L is 50, C is 100, D is 500 and M is 1,000). Using Roman numerals, the number 365 is CCCLXV, but there is no hope of making a tidy anagram out of that, for there aren't any vowels to play with. But then God is a very mighty and strange spirit. Did it not occur to the Gnostics that he might really be called Ccclxv? In which case they would no longer need to search for his name; only a means to pronounce it.

History suggests to us that the Gnostics were not impressed by the Roman solution. They hated the Romans and, in any case, the 22 letters of the Hebraic alphabet had already been allotted numbers by Hassidic rabbis and other Cabbalists trying to interpret the scriptures. The Gnostics naturally turned to these.

Incidentally, the word 'cabal' which is nowadays opprobriously used to mean a secret clique or faction, comes originally from the Cabbala, an ancient Jewish tradition (of which the Gnostics formed the most forceful faction) which interpreted the Bible using mystical ciphers. In seventeenth-century England, Charles II's C.A.B.A.L. was a derisive term, used by the press to describe his Cabinet, whose members' names (Clifford, Arlington, Buckingham, Ashley-Cooper and Lauderdale) provided the acronym.

The ancient art of turning letters into mystical numeric ciphers had been practised by the Egyptians several thousand years before the Gnostics came on the scene. In this woolly headed scheme, words were given numbers often for the silliest reasons. The human head, for instance, was given the number seven, because it contained seven orifices – two ears, two nostrils, one mouth and two eyes. The ibis was number eight because it was the incarnation of the god Thot, who was chief overlord of the city of Khemenu, which means 'the city of eight', and in a very early example of what seems to be cockney rhyming slang the number nine is encrypted as a sun with descendent rays, meaning 'shine' – both 'nine' and 'shine' in Egyptian are pronounced 'psd'!

So as far as the Gnostic was concerned, 365 was his super sod ('sod' is the Hebrew word for 'secret'). By taking the numbers from the Hebraic alphabet there were many different ways of coming up with that sod – 365. For instance, the seventeenth letter of the Hebraic alphabet, 'PE' (pronounced 'P') is worth 120, so let's have three of those – that's 360, and then we need another five. Let's take five ALEPHs (that's the first letter of the alphabet, worth one). Now what do we have? AAAAAPPP? Worth 365, but, alas, a silly name for God.

'I know, let's swap two of the ALEPHs for one BET, which is worth two. Now what have we? AAABPPP. Mix it around. It's bound to be an anagram – BAPAPAP. That's it! God's name must be BAPAPAP. Hooray! We have found the *sod*! Oh mighty BAPAPAP.'

'And look, BAPAPAP, you've got seven letters. That's magic too. We have 365 and seven, all in one. We've got it!'

'Hang on, you lot. How do we know it is not PABAPAP, or PAPABAP, or even . . . ?'

'I think the *sod* is ABRASAX.'

'What?!??!'

'ABRASAX has seven letters too, and they all add up to 365. ALEPH is one, BET is two, RESH is 100, another ALEPH – that's one – SAMEKH is 200, one more Aleph and a Quf, which is 60. It all adds up to 365. How about it? ABRASAX?'

The African Bishop – later St – Cyprien (200–258 AD) had another, very sensible reason to assert 365 as the magic safe-breaking clue to God's name. Cyprien's reasoning had nothing to do with the number of days in the year. It went like this: T (the symbol of the cross) is 300; YH (the first two letters of the tetragrammaton – explained in a minute) is 18; add 16 (being the number of years into the reign of Tiberius when Jesus was crucified); then add another 31 (Jesus's age, according to Cyprien, at the time of his crucifixion). Put all these together and you get 365 – which *proves* it's a special number!

The futility of this practice needs no further elucidation. The possibilities were endless. They had not even decided on what language God's name was supposed to be in, let alone which number was the magic number. Over 1,000 years were spent in this idle pursuit. In the end some of them settled for ABRASAX, engraving the name on gems that could be used as amulets or talismans; others rebelled, deciding that the magic number, 365, was wrong. It should be 284, for this was the number of three important Greek words: *Theos* ('God'), *Hagios* ('holy') and *Agathos* ('good'). The number 17 was likewise venerated because it spelt *Tov*, Hebrew for 'good'. By the same judicious methods the Gnostics can claim to be the proud inventors of the childrens' party word Abracadabra, supposedly made up of the Hebrew words *Ab* (Father), *Ben* (Son) and *Ruach ACadsch* (Holy Spirit).

If this is so, it is hard to understand why 'Abenrucad' would not have worked just as well. In the Middle Ages Abracadabra was written as a triangular cipher on a piece of parchment, tied to a strip of linen and dangled, like a necklace, around any person suffering from flu, gout, toothache, palsy, the ague or the stitch.

Over the whole issue of God's name, the Bible (for ever the muddler in these matters) is perhaps partly to blame for the mess the Gnostics found themselves in. To Moses God says: 'I AM THAT I AM; Thus shalt thou say unto the Children of Israel, I AM hath sent me unto you.' And a few chapters later in Exodus God says: 'I appeared unto Abraham, unto Isaac, and unto Jacob, by the name of God Almighty, but by my name JEHOVAH was I not known to them.' In the Book of Psalms we are told to 'Sing unto God, sing praises to his name: extol him that rideth upon the heavens by his name JAH, and rejoice before him.'

To the Jews, God's name was symbolised by four special letters – YHWH – known as the tetragrammaton. They never spoke these letters, for they were considered too sacred to utter. What they mean is 'I AM', and these letters, fleshed out, can be pronounced as 'Jehovah'. It is a name which supposedly gives to God all the most mystical characteristics which the scriptures apply to him. 'I am' means he is the alpha and the omega, the beginning and the end; the past, the present and the future, the 'whole being', as reflected in Jesus's answer to the Jews in the Gospel of St John: 'The Jews said unto Jesus, Thou art not yet fifty years old and has thou seen Abraham? Jesus said unto them, Verily, verily I say unto you; Before Abraham was, I am.'

Brewer, in his famous dictionary, remarks how odd it is that, in so many languages, the name for the supreme being should be composed of four letters:

Thus there are in Greek *Zeus* and *Theos* (spelt in Greek with a single theta for the 'th'), in Latin *Jove* and *Deus*; French *Dieu*, Dutch *Gotd*, German *Gott*, Danish *Godh*, Swedish

Goth, Arab *Alla*, Sanskrit *Deva*, Spanish *Dios*, Scandinavian *Odin* and our *Lord*.

Did the Gnostics not consider four as a supremely magic number? To Pythagoras it was perfection.

Nowadays we are fearless. Cynicism has set in too deep for us to fret about the meaning of numbers, which is perhaps just as well when you consider all the time-wasting that our credulous ancestors got up to. We are a more refined breed. Nowadays we cannot muster interest in practically any number, unless, of course, it is peppered with noughts, as many as possible – yes, the more of those the merrier. A tenth anniversary is an excuse for a glass of champagne; the passing of a century warrants a case; and a new millennium means that France runs dry of the stuff. A simple equation sums up our extraordinary modern exuberance at the mere suggestion of decimal numbers: $P=ch^2$ (where P is the number of zeros tacked on to the end of a numeral and ch is the quantity of champagne accordingly drunk).

The ancients talked about 'uncertain numbers', by which they meant 'round numbers'. In Roman numerology there is no circular symbol but, as the saying goes: *Numerus certus pro incerto ponitur* ('We use a specific number to denote an uncertain one'). We do the same today: 'A thousand wasps flew out of the nest' does not mean exactly 1,000 wasps but just lots of wasps – too many to count. As Byron says in the third canto of *Don Juan*:

> But words are things, and a small drop of ink,
>   Falling like dew upon a thought, produces
> That which makes thousands, perhaps millions, think;
>   'Tis strange the shortest letter which man uses
> Instead of speech may form a lasting link
>   Of ages; to what straits old Time reduces
> Frail man, when paper – even a rag like this—
> Survives himself, his tomb, and all that's his.

And in Shakespeare's *Merchant of Venice* we are told: 'Three thousand ducats; 'tis a good round sum.'

There is only one zeroless number which interests us: 666. Known as the number of the Beast and referred to in the Revelation of St John: 'Let him that hath understanding count the number of the beast; for it is the number of a man; and his number is six hundred, three score and six.' Hollywood has milked this remark for all it is worth. It has given us little boys called Damien and demonic dogs with the number tattooed under their hair; it has given us 666, the phone number for hell as well as Satan's car numberplate. But we need not be afeared. No Antichrist is lurking around the corner with this number etched on his bottom, nor is 666 any more scary or interesting than any other number under the sun. All we can infer from it is that St John was another wretched Gnostic. If you spin the words of Nero Caesar around in Hebrew you can find the number 666 by adding the values of letters together. (Nero, of course, was a great persecutor of the Christians.) Using the devious processes employed by the Gnostics, you can make the 666 label stick to just about any figure in history. The emperors Trajan and Hadrian, for instance, have been denounced as the Beast. So too have the Emperor Diocletian; Julian the Apostate (half-brother of Constantine the Great) who declared himself a pagan; Muhammad the Prophet; Camillo Borghese (known as Pope Paul v) who had the effrontery to excommunicate the Doge of Venice; Silvester ii, a learned and musical tenth-century pope; Napoleon Bonaparte; Charles Bradlaugh, an English agnostic, reformer and iconoclast; Wilhelm ii of Germany (more on him shortly); and Margaret Thatcher, Prime Minister of Great Britain throughout the 1980s. There is one other, milder suggestion as to why St John might have chosen to vilify the Beast as number 666: namely that seven is the holy number of the Jews and 666 falls short of it three times. It seems strange, though, that a number so close to perfection should be demonised as the Beast, for surely, if this were the case, then 333 or 222 would be considered even more evil than 666.

So much for the Beast. The number 100 has no sacred overtones; it is neither the number of the devil nor the number of our Lord, and yet we plainly adore it. We adore it because of its lovely noughts and because of the relationship it enjoys with 10, with 1,000, and with 1,000,000. We weigh and we measure in hundreds; most of the world's currencies are based on the decimal system. It is a small wonder that only a few of our units of time-measurement are based on the radix 10 – the microsecond, the decade, the century and the millennium. As an interesting aside, it is also worth noting that, in English, all numerals and ordinals from zero to one million are derived from Anglo-Saxon words. There are two exceptions to this rule: 'second' comes from the French (the Anglo-Saxon word being 'other', which is clearly confusing in a sequence 'first, other, third'); the second exception is the word 'million', which derives from the Latin *mille* – 1,000.

In the early days, that is to say before the eighteenth century, the word 'century' was most typically used to apply to any collection of 100 things (not necessarily years). As in Imogen's beautiful, lachrymose speech from Shakespeare's *Cymbeline*:

> I'll follow sir, But first an't please the gods,
> I'll hide my master from the flies, as deep
> As these poor pickaxes can dig; and when
> With wild wood-leaves and weeds I ha' strew'd his grave,
> And on it said a century of prayers,
> Such as I can, twice o'er, I'll weep and sigh
> And, leaving so his service, follow you.

Just as the word 'decade' was extracted from the phrase 'a decade of years' so a 'century' was likewise first and foremost a century of years. In cricket, a game replete with arcane jargon, the word is still used to define 100 runs as scored by a single batsman. In the Roman army a *century* was a unit of 100 men commanded by a *centurion*. In each and every case they bow to the glory of the decimal system. From the early part of the seventeenth century there are only a few extant examples showing the word used to refer to specific centuries dating from a chronological epoch. Not until the eighteenth century and, more generally, the nineteenth century did people start talking about centuries with specific names. Most people in the eighteenth century were unaware of exactly which century they were living in and, as with decades, people's understanding of dates, as we have seen, was far more anchored to the regnal years of kings and queens. It was not until the last decade of the nineteenth century that everyone, in Europe at least, knew where they were in the Dionysian epoch and where they stood on a timescale in relation to the birth of Christ.

The speed and advancement of technology has been a recurrent theme of people reflecting back over a dead century. To the Victorians of the British Empire it was important to take a pride not only in the empire itself and the mother nation, but in the time and century as well. Queen Victoria's Diamond Jubilee was celebrated in 1897, providing the British with an excuse for an explosion of jingoism. But the same was happening right across Europe and America too. Jingoism was not just about geography and culture; it was also about time. For the first time in history it was the century that mattered. To be born in the right country and during the right century was the high privilege of the nineteenth-century jingoist. So fervent was his patriotism and satisfaction with the time and place in which he found himself that it became fashionable in more chic metropolitan circles to look the opposite way. In *The Mikado*, an operetta by Gilbert and Sullivan satirising the whole stratum of English society, in Japanese costume, Ko-Ko denounces certain types in his famous catalogue song 'I've got a little list'. Among his pet hates are 'all people who have flabby hands and irritating laughs, people who eat peppermint and puff it in your face and the idiot who praises, with enthusiastic tone, all centuries but this and every country but his own'.

In 1898 Alfred Russel Wallace, the English biologist (who, with Darwin, promoted the evolutionary theory of natural selection), summed up the mood in his proud review *The Wonderful Century*:

Not only is our century superior to any that have gone before it, but . . . it may be compared with the whole preceding historic period. It must therefore be held to constitute the beginning of a new era of human progress.

Two years later, the century came to its close. For the first time in history everybody – the rich and poor, the stupid and the bright – were aware of the date, and the significance of the passing of the century. It gave pause for thought, both to reflect on the passing years and to anticipate the coming era. *Popular Science Magazine* of 1898 balked at the terrifying acceleration in technology which had, according to it, been the defining characteristic of the nineteenth century:

> As the nineteenth century draws to its close there is no slackening in that onward march of scientific discovery and invention which has been its chief characteristic. At the beginning of the century the telegraph was as yet undreamed of and the telephone and the dynamo utterly unimaginable developments. Had anyone dared to conceive that signals could be made to pass in a second of time between Europe and America he would have been considered a fit candidate for Bedlam.

One hundred years earlier the same sentiment was reported in the introduction to the Annual Register of 1800:

> On the general recollection or review of the state of society or human nature in the eighteenth century, the ideas that recur oftenest, and remain uppermost in the mind, are the three following: the intercourses of man were more extensive than at any former period with which we are acquainted; the progression of knowledge was more rapid, and the discoveries of philosophy were applied more than they had been before to practical purposes. This present age, in respect of former times, may be called an age of humanity. Whence this happy change? Not from the progressive effects of moral disquisitions and lectures . . . but

from the progressive intercourses of men with men, minds with minds, of navigation, commerce, arts and sciences.

If this is the case for the eighteenth and nineteenth centuries, the twentieth was, more than any other, the century of technological progress; rapid worldwide communications, more liberal government and, of course, computers. But it was also the era of dehumanisation, where machines took over jobs, television replaced the traditional social hours of eating and talking, and power was spread among committees and boards. Writing for the *New York Times* in 1969, Russell Baker commented: 'At its mid-afternoon the twentieth century seems afflicted by a gigantic and progressive power failure. Powerlessness and the sense of the powerlessness may be the environmental disease of the age.'

It is a sallow judgement, voiced as early as 1925 by T. S. Eliot in his gravely pessimistic poem, 'The Hollow Men', in which the poet views modern society as 'shape without form, shade without colour', and describes the terrifying emphasis of modern life as a 'paralysed force, gesture without motion'.

As far as time and space are concerned, the twentieth century brought with it deeper understandings, but in terms of the exponential advances in communications and other such technologies it ended with a surprising number of questions still unanswered. On a personal level time and space seemed to be shrinking, as was the size of the world itself, in the gormless face of ever-expanding technology. In 1938 the British Prime Minister was able to describe Czechoslovakia as a 'far-away country'. By the end of the century Czechoslovakia no longer existed, and Britons were buying books over the Internet from shops in Australia, 12,000 miles away. As one writer, Edwin Way Teale, reflected, as early as 1956: 'Time and space – time to be alone and space to move about – these may well become the great scarcities of tomorrow.'

And so the century, any century, is nothing but a polymorphous mass of distantly related events, bracketed together in celebration of our love for the number 100. What would we think of the century had it lain between 1842 and 1942? Or what about the one that will run from 1994 to 2094? Nothing. We will never comprehend or acknowledge such arbitrary divisions. We shall simply refuse to recognise them. They may be centuries of 100 years, but not the right 100 years – where are all the noughts? We cannot think historically without noughts to guide us in our views. Henry Wallace (1888–1965), President Roosevelt's Minister for Agriculture, was speaking prophetically when he wrote in 1942: 'The century on which we are entering – the century which will come out of this war – can be and must be the century of the common man.' The common man may have gained the seats of power, but the century from 1945 to 2045 will never be recognised.

'*Soldats, songez*,' ordered Napoleon to his troops before the Battle of the Pyramids in 1798, '*que, du haut de ces pyramides, quarante siècles vous contemplent*.' ('Imagine, soldiers; from the top of those pyramids forty centuries look down on you.') There is something poetic in these words which may have spurred his troops to victory. For Napoleon and his troops, it is the centuries themselves that are looking down on them, not the pyramids – and that means all those precious noughts. For the century imprints its own romantic image on to history and so, as we have seen, does the decade. They are the frames that change the pictures. Our view of the past is skewed by our fascination for numbers, and so is our view of the future too. In many ways we are no more enlightened than the fatuous Gnostic with his panting joy at the thought of 365, the scaremonger St John with his 666, the Romans worried to bits by the number two or the Egyptians beaming with pleasure at 360.

It is a fluke that it has all happened this way, a fluke that Dionysius Exiguus gave us a year for the birth of Christ and told us to start counting from then; a fluke that we obeyed him; a fluke that we evolved with just so many fingers; a fluke that we are alive at all. How different would history have been if we had sorted our calendar by a different radix? If we hadn't been warped by our love for the number 10? Look at it through the eyes of that 'funny thing', that hatted, hairy, benign-looking creature from the children's beginner book, *One Fish, Two Fish, Red Fish, Blue Fish*:

> Say!
> Look at his fingers!
> One, two, three . . .
> How many fingers do I see?
>
> One, two, three, four,
> five, six, seven,
> eight, nine, ten.
> He has eleven!
>
> Eleven!
> This is something new.
> I wish I had
> eleven, too!

And if Napoleon had been one of these 'funny things' – what would he have said to his troops then, eh?

Chapter XI

MILLENNIA
Millennia

Acryptic little comment which appears in Chapter 3 of the Second Epistle of Peter in the New Testament has caused a huge rumpus which is still brewing today: 'But beloved,' Peter quotes as the words of David, 'be not ignorant of this one thing, that one day is with the Lord as a thousand years, and a thousand years as one day.' It seems innocent enough. After all, it has long been held that time, as it is understood by God, cannot be the same as time seen from the point of view of us mere mortals. After all, God was supposed to have created time or (if you take St Augustine's line on this) forged the universe out of time itself. If this is so, we can hardly expect God's idea of time to be the same as ours. Was David being precise when he said that 1,000 of God's years are as one to us? Or was he just bandying the number 1,000 because, like all people who live in a decimal culture, it is a vague term often used to mean 'a very large number'? That may be so, but adamant theologians will point to David's opening words: 'be not ignorant of this one thing'. Is this how people normally preface an off-the-cuff remark?

Maybe David was being exact. But so what? What is the big significance? 'Ah ha!' responds the wily theologian. 'If God created the world in six days and each of God's days is, to us, the same as 1,000 years, then the whole process of creation, which is recorded in the Bible as having taken place in six days, must have lasted (in our terms) 6,000 years.'

At some point the 'must have lasted' part of that argument got changed into 'must last'. For it was generally believed that the world could not possibly be more than 6,000 years old, and if it were less than 6,000 years old and David's point about one day in God's time being the same as 1,000 years in ours, then what? – 'Alarmy! Alarmy!'– the Creation must still be going on. And if the Creation is still in progress, where on earth does that leave us?

Back to the scriptures:

'And God said, Let there be Light.' Yes, yes, we know that; so that's been done, obviously: sun, moon, yes, done that; where are we now: three days? 'And God created great whales and every creature that moveth.' Yes, good. 'I have given you every herb-bearing seed which is upon the face of all the earth.' What's next? Quick, I must know.

If you detect a note of panic in the scholar's tone, it is because he is scared, scared out of his wits that the Last Judgement is about to happen. All sinners (which is most people) will be found wanting and shoved into the pit of the earth to serve an eternal ordeal of boiling sulphur and bubbling waist-high excrement. But why should the Last Judgement be about to happen?

The scare story is a simple one. It runs like this: God created the world in six days (implying 6,000 human years) and on the seventh day he rested. The sixth millennium, corresponding to the sixth day of the Creation (in which God made Adam and Eve), is therefore the time at which we can expect Jesus to return to earth in human form to have a massive fight with the Prince of Darkness, to vanquish him for ever, and to lead his believers and repenters through the great hall into his father's dining room for a feast of grapes, grain, milk, honey and 'fat things' (we shall deal with the milk, the honey and the fat things later on).

Let the eighteenth-century English cleric, Thomas Burnet, explain:

It is necessary to show how the Fathers grounded this comparison of six thousand years upon scripture. 'Twas chiefly upon the Hexameron, or the Creation finished in six days, and the Sabbath ensuing. The Sabbath, they said, was a symbol of the Millennium, that was to follow at the end of the world; and then by analogy and consequence, the six days preceding the Sabbath must note the space and duration of the world. If therefore they could discover how much a day is reckoned for, in this mystical computation, the sum of the six days would be easily found out. And they think, that according to the Psalmist and St Peter, a day may be estimated a thousand years; and consequently six days must be counted six thousand years for the duration of the world. This is their interpretation, and their inference.

Burnet, by the way, was a curious figure. He believed that hell was a place from where the damned could be returned if they repented their sins. This view, which is contrary to normal Christian teaching (which states that once a sinner has entered the gates of hell he will remain there for the rest of time), set Burnet apart in his day. Many accused him of heresy, as one contemporary commented: 'Burnet, like a clever seducer, advises that one should proceed step by step, so that his new doctrine may, like a cancer, make imperceptible progress.'

In this instance, though, Thomas Burnet was simply describing the phenomenon of growing millennialism, a branch of Christian thinking which fervently believed that the seventh day, on which God took his rest from the labours of the Creation, would be represented by the seventh millennium after the Creation – a time of everlasting peace, for the meek at least.

There is something deeply implausible about this whole panic-stricken line of thinking. For a start, if God's days were really the same as 1,000 years to us (i.e. 365,000 days to us), then presumably our days would likewise be 365,000 times shorter when they are experienced by God. When the Book of Genesis says that God created the world in six days it is not stated whether those six days are 'human days' or 'God days', which, as we have seen, differ in time-duration by a factor of 365,000. If Genesis meant to say that God created the world in six 'God days', then fine, we can, if we wish to stretch our imaginations beyond the horizon and back, interpret this as meaning that he created the world (or intends to create the world) in 6,000 'human years'. If, on the other hand, what Genesis is trying to say (and seems to be succeeding in saying quite clearly) is that God created the world in six 'human days', then as far as God is concerned the act was accomplished in 1.4202 of a second – hardly long enough to justify a rest on the seventh day, especially since his rest would have lasted no more than 0.2367 of a second. If Genesis was talking about 'human days', then the whole millennium scare is a complete waste of time. But even if God had created the world in six 'God days', the structure of the argument remains hopelessly dodgy.

The opening two chapters of the Book of Genesis make it abundantly clear that the Creation has been and gone. It is written in the past tense: 'I have given every green herb for meat,' says God, and earlier, 'And God made the beasts of the earth.' Well, we know that whales exist already, and we know that all the herbs for meat exist, we know that light and the sun and moon have been created already, so why do these obstinate millennialists suddenly decide that the whole thing is figurative and that the Creation is still in progress?

Part of their worry stems from the conflict of having a God who is supposed to be infinitely kind and good and wise and pleasant yet who has created a world that we can all clearly perceive to be full of odium, disease, vicious smells and evil. They need an excuse for this sorry state of affairs: 'Oh, yes,' says the millennialist apologist, 'but, you see, he hasn't actually finished creating the universe yet. In fact, he's hardly started. This is just a test run, in which the devil has a little patronage here and there, but the real thing lies beyond. He will destroy the Antichrist on the sixth day. It will come to us in the seventh millennium after the Creation, when all the evil has been vanquished from this world.'

So if people really wanted to know when the Apocalypse was

going to happen they had, first of all, to work out when the Creation took place and, from there, count forwards 6,000 years, and – bingo! – you can calculate a date for the end of the world.

These untamed shreds of crude hyperbolic reasoning were to prove the beginnings of an extraordinarily illogical process which would occupy some of the brightest minds in the history of human civilisation for several thousand years.

So if the Creation of the universe has not yet ended, at what stage are we presently at in this 6,000-year process?

In order to calculate a date for the beginning of time, scholars needed something to hook on to, or the whole thing was too nebulous. They could count generations and eras backwards using the genealogies of the Bible and ancient histories, but this was a fishy business which involved an uncomfortable degree of fudging. There must be something more concrete than that. The obvious answer (for the Christians at least) was to bring Jesus into the picture. He was, after all, an historical figure. If Jesus was the son of God, then obviously God would not have dumped his son on earth at any random point in his creation scheme. That would have been too undignified.

'No. God was a wise being, and wise beings do not do that sort of thing. A wise being would always make sure that his son (who is also himself in this instance) arrived on earth, not *kerplonk* in the middle of a millennium, but on the join, just at the point where one millennium ends and the next one begins.

'Good, so we have established that! Jesus was born on a millennium join. All we have to do now is ascertain *which* millennium join he was born on, and then, by counting backwards, we shall know exactly when the Creation began, and by counting forwards we shall be able to foresee when the world is due to end too.'

So there was a problem for the theologians to squabble over, and squabble they did. If Jesus had been born on one of the 1,000-year joins (and that was now more or less decided), was the first day of the Creation 1,000, 2,000, 3,000, 4,000 or even 5,000 years before his birth? Scriptural scholars, right up until the late eighteenth century, were all agreed that God must have created the world on one of these dates. The difficulty was to figure out which.

You may remember Archbishop Ussher, the seventeeth-century Irish theologian, who published a saying that the Creation took place in 4004 BC. This is how he put it:

> The true nativity of the saviour was a full four years before the beginning of the vulgar Christian era, as is demonstrable by the time of Herod's death. For according to our account the building of Solomon's Temple was finished in the 3,000 year of the World, and in the 4,000 year of the World, the days being filled in which the Blessed Virgin, Mother of God, was to bring forth Christ himself, was manifest in the flesh, and made his first appearance unto man: from which four years being added to the Christian era, and as many taken away from the years before it, instead of the Common and Vulgar, we shall obtain a true and natural Epocha of the Nativity of Christ.

But there was also another, smaller problem to solve, and it needed their attention first. Before deciding which millennium join Jesus was born on, it would be just as well to sort out, once and for all, when Christ was actually born, according to the existing calendar.

The generally accepted view is that Jesus's birth took place four years 'before Christ' – a strange irony indeed, but one that is based on the historical and biblical wisdom which states that Christ was born in the reign of Herod the Great, a king who was known to have died in 4 BC. It is well documented that Herod was Governor of Galilee from 47 BC. His father Antipas (not to be confused with his licentious son – also called Antipas – who beheaded John the Baptist for the thrill of seeing his stepdaughter, Salome, strip) was given the governorship of Judaea by Julius Caesar, and Herod the Great subsequently stabilised his own position at Galilee with frequent trips to Rome for the purposes of endearing himself to Roman high society. He was on particularly friendly terms with Caesar's cousin (and Augustus's brother-in-law), Mark Antony, though when the tide turned, Augustus had a fallout with Mark Antony and Herod forged an even closer bond with the emperor. His sucking up was such a success that Augustus made him King of Judaea in 40 BC, a position he held for 36 years until his death. He committed many vile crimes; he murdered more than one of his sons and, in fear of the prophesy that Jesus would become King of the Jews, dispatched the Wise Men to find the baby in the hope that once he knew Jesus's whereabouts he would have him murdered. When the Wise Men failed to produce, Herod issued his famous decree ordering the assassination of all male children of two years or under – a hideous act that would nowadays be classified as a 'crime against humanity'. The slaughter of the innocents has been captured with particular force by the painters Pieter Bruegel the Elder and Guido Reni. A version in white marble relief on the floor of the Duomo in Siena has been a source of constant fascination for over 400 years.

Herod's death is confirmed by a Jewish historian, Flavius Josephus, and the account of his dying shortly after an eclipse and before the feast of the Passover allows us to place his last breath on earth, with some degree of confidence, at between 13 and 30 March 4 BC. This means that if any part of the story concerning the Wise Men and Jesus is true, then Jesus must have been born either in or sometime before Herod's death in 4 BC. If Herod slaughtered the innocents in 4 BC and had reason to believe that the young Jesus was aged two or less, then Jesus might well have been born in 5 or 6 BC. If Herod slaughtered the innocents before 4 BC, then who knows when Jesus could have been born?

Dionysius Exiguus (a figure who will loom larger in the pages to follow) is regarded as the brains behind our modern BC/AD system. His incorrect placing of Jesus's birth in 1 BC was based on an old text which stated that Christ was born in the twenty-eighth year of the reign of Emperor Augustus. We might forgive him for a certain amount of confusion here, since Augustus was accepted to the title of emperor at a ceremony in Rome on 13 January 27 BC. However, Augustus (or Octavian, as he then was) was recognised as the undisputed master of the triumverate four years earlier, when he defeated Antony and Cleopatra at the Battle of Actium on 2 September 31 BC. It is almost certain that the text by Clement of Alexandria, which Dionysius was relying on for the nativity date, was, in fact, alluding to the earlier date as the start of Augustus's reign. In which case the twenty-eighth year would not be 1 BC but 4 BC, the date favoured by Ussher.

So let's, for the sake of proceeding forward with this argument, accept the fact that Jesus was born in 4 BC. Now if Jesus's birth marked the beginning of the fifth phase of the Creation (which, with a splash of Ussher's sophistry, it did) then, by counting backwards (using the one day = 1,000 years rule), Ussher and his friends arrived at the date of 4000 BC, adding four extra years to straighten out the birth-date irregularity and finally settling on 4004 BC for the date of the Creation.

But Ussher's calculations were not altogether right. If they had been, then the mighty Apocalypse which he prophesied should have occurred 6,000 years after 4004 BC in AD 1997. Whoops! – he got it wrong. But Ussher's hand was forced by the strictures of his own fatuous millennium theory. He could not say that the Creation happened in 3004 BC because there was evidence that human civilisation was around before that; nor could he claim that it had taken place in 5004 BC because that would have put the date for the Apocalypse 6,000 years later, to AD 997, and he knew that nothing so dramatic as the end of the world had happened on that date – he would not have been around to spout his millennium rubbish if it had.

At the turn of the first millennium, a Flemish monk, Roual Glaber, had predicted dreadful things: 'The Prince of Darkness will soon be upon us because the thousand years have been completed.' He had decided the sixth day (the sixth millennium) was represented by Jesus's first (and so far only) visitation on earth, which meant that the dreaded seventh millennium was due to start from midnight of 31 December AD 999. When the millennium arrived and nothing happened, Glaber changed the date of expectation to 1033 (i.e. moving the date from Christ's birth to the millennial anniversary of his crucifixion). When 1033 came around he must have noticed he was wrong again. 'There is nothing for it but to collapse in deepest humiliation,' as the scientist Arthur Eddington would have said.

As more and more millennial predictions turned out to be wrong, wary Christians realised that more moving of the goalposts was probably needed.

If the historian's ingratitude to Dionysius Exiguus for getting the birth of Christ wrong is pitiful, it is as nothing compared with the futile and often ill-mannered debate that has raged for 1,500 years, and still rages now, over his system for numbering the years.

What he did to so annoy the pedants was to call the first year of his new system AD 1 instead of calling it zero. We will come to why he called it AD 1 in a moment, but another question first: why do the bores and pedants of this world think that he should have called it zero? As a correspondent to a London newspaper points out:

> There can be no such thing as a year zero, or, for that matter, a second zero, or an inch zero, or any other unit zero, since zero is a point with no dimensions, separating plus units from minus units on any normal scale (try finding it on a logarithmic one).
>
> Thus, in the Christian calendar, zero is the point in time that was (when and by whom and how inaccurately is irrelevant) ascribed to the birth of Christ. The previous year, which we call 1 BC (the first year Before Christ), is therefore minus one on our timescale in years, and the year following the zero point, which we call AD 1 (the first year of Our Lord – Anno Domini), is plus one.

The argument about when to celebrate a new century was already in full flow in the 1690s. In the 1890s it had come to such a head that decisive action needed to be taken. The German Emperor Wilhelm II was known to be a political vacillator, always drifting with the flow of things and never standing up against them. He was not what you would call a leader in the heroic mould. He had a damaged left arm that never grew to its full size, which some historians have cruelly cited as the cause for his other deficiencies. When he gave an interview to the *Daily Telegraph* in 1908 he tactlessly let slip that most Germans were anti-British. The remark caused ripples of indignation to be felt in and around London, but by and large he was not a figure to be noticed. As the grandson of Queen Victoria, his own anti-British stance might have seemed a little disloyal, until it is remembered that the British queen herself was German. Even so, Wilhelm's order from Aix-la-Chapelle, dated 19 August 1914, which found itself held to ridicule in the London *Times*, shows a touch of hysteria on the matter:

> It is my Royal and Imperial Command that you concentrate your energies, for the immediate present, upon one single purpose, and that is that you address all your skill and all the valour of my soldiers to exterminate first the treacherous English, and to walk over General French's contemptible little Army.

In November 1918 he was forced to abdicate his throne after the defeat of the Germans in the First World War, and for the rest of his life he loafed about as a country squire at his estate in Holland, until he died in 1941. It was perhaps because of his reputation for political inertia that he tried, when emperor, to market himself as a great decision-maker. In that spirit he announced to the world that the new century was officially to begin on 1 January 1900, proclaiming that the world was now 'facing a still brighter dawn of civilisation'. On 1 January 1900 he ordered a 33-gun salute to the new century.

Not everybody was so impressed, for while Wilhelm had, once again, taken the people's line (most people were planning to party on New Year's Eve 1899 anyway), there were still, particularly in the academic professions, many people determined that 1 January 1901 must be commemorated as the true first day of the twentieth century. The popular press was supporting 1900, and people, as a whole, followed their lead.

In New Jersey a German-American couple arranged to be married at 12.01 a.m. in the hope of being the first couple to be married in the twentieth century. Perhaps they had never heard of the International Date Line. The whole of Europe, Africa, Asia and Australasia lie ahead of New Jersey in Standard Time – so they were probably among the last couples to be married on 1 January 1900. In any case the debate was far from settled. A century earlier, an unknown poet fuelled the argument with this charming piece of doggerel published in the *Connecticut Courant* on 1 January 1801:

> Precisely twelve o'clock last night,
> The Eighteenth Century took its flight.
> Full many a calculating head
> Has rack'd its brain, its ink has shed,
> To prove by metaphysics fine
> A hundred means but ninety-nine;
> While at their wisdom others wonder'd
> But took one more to make a hundred.

The number of seemingly bright people who completely fail to understand this simple problem lends a fascinating angle to the study of the human brain. A Harvard professor, zoologist and well-known writer, Stephen Jay Gould, an unquestionably clever man who should, on no account, be associated with the pedants and bores above, is uncertain. Even in his special study of the problem, a collection of essays called *Questioning the Millennium*, he finds himself in a terrible muddle:

He [Dionysius Exiguus] started time again on January 1st,

754 A.U.C. – and he called this date January 1st of year one A.D. – not year zero . . . In short, Dennis neglected to begin time at zero, thus discombobulating all our usual notions of counting. During the year that Jesus was one year old, the time system that supposedly started with his birth was two years old. (Babies are zero years old until their first birthday; modern time was already one year old at its inception.)

The roots of this misunderstanding lie deep, and if we are to avoid too many interjections of 'Yes, but . . .' we must unravel it with care.

Firstly Dionysius did not 'forget to begin time at zero'. In point of fact he did exactly what Stephen Jay Gould and others like him wanted him to do. He made AD 1 the First Year of Our Lord, which included Jesus's *first* birthday on 25 December AD 1. There should be no problem with this. If someone is born in 1960 they would be one year old in 1961 and would be 10 in 1970, celebrating their first decade of life. It is exactly the same with Jesus. The Dionysian era began on 1 January AD 1. Jesus was born, according to Dionysius, on the eighth day before (i.e. 25 December 1 BC). So 1 January AD 1 therefore represents, not the birth of Jesus, but his eighth day of life, the feast of the Circumcision. Let the great Moses explain this one:

This is my covenant, which ye shall keep between me and you and thy seed after thee; Every man child among you shall be circumcised. And ye shall circumcise the flesh of your foreskin; and it shall be a token of the covenant betwixt me and you.

And he that is eight days old shall be circumcised among you, every man child in your generations, he that is born in the house, or bought with money of any stranger, which is not of thy seed.

He that is born in thy house, and he that is bought with

thy money, must needs be circumcised: and my covenant shall be in your flesh for an everlasting covenant.

And the uncircumcised man child whose flesh of his foreskin is not circumcised, that soul shall be cut off from his people; he hath broken my covenant.

If you remember that Jesus was never, even by Dionysius, supposed to have been born in AD 1, then things become a little bit clearer. Where the arguments start is over the issue of when to celebrate centenaries and millennia. The silliest of all arguments runs roughly like this: 'A baby is deemed to be "one" after his first birthday and to be "two" after his second birthday (i.e. after he has lived for two full years from the date of his birth). Therefore, on this basis, if time was started on 1 January in any given year, then time (like a baby) should be "one" on 1 January of the following year. However, in the Dionysian scheme of things, it is apparently two, because the name of the year is AD 2, not AD 1.'

In a barely more sensible codicil to this line of thinking, the pedant continues: 'A decade is supposed to be a period of 10 years. We should not celebrate a millennium (say, AD 2000) on 1 January 2000 because on that date only 1,999 years have passed since 1 January AD 1. The two-thousandth year is only completed at midnight on 31 December AD 2000, and not until 1 January 2001 can we truthfully claim to have entered the third millennium. On this basis the nineteenth century started on 1 January 1802 and ended at midnight on 31 December 1901.'

These carping fusspots, having made their point over and over again, inevitably fail to observe their own teaching. With their party hats on, flushed with the merriment of wine and lamely miming the conga, they were to be seen on New Year's Eve 1999 just like everyone else:

'Oh, ho, ho! We really shouldn't be doing this, you know. Really, it's not right, it isn't. The third millennium isn't due to commence for a year.'

'Is that so? Well, I never did!'

'Honestly. Look, if the whole thing started in AD 1, then it stands to reason, doesn't it? The third millennium begins on 1 January *next* year.'

'Oh, what the hell. Have another!'

Of course, it is true that a decade should hold 10 years in it and that a millennium is a unit of 1,000 years, but who ever said that the celebrations on 1 January 2000 were, or ought to have been, held in honour of the completion of 2,000 whole years since 1 January AD 1? For most people the celebration of a millennium is simply a question of numerical excitement. Two thousand is, after all, a very neat number with all those noughts after it, and we just love the idea of a stodgy old set like 1999 changing *all four* of its digits into a minty new set suddenly overnight. As a correspondent to the *Daily Telegraph* in January 1999 observed:

> The apparent impressiveness of next year's date [2000] occurs simply because we are using a number system based on 10, the decimal system. If we were using the duodecimal system, with a radix of 12, then the date of 2000 would become 11A8, where the 'A' replaced the decimal 10.
>
> The same date could also be denoted as 3720 in the octal system and 11111010000 in binary.
>
> So what is all the fuss about?

The other less popular reason we (or Christians at least) celebrate a millennium is to honour the birth of Christ. Let us not forget that Christ, according to Dionysius, was born on 25 December 1 BC (not, as people often suppose to be the case, on 25 December AD 1). This means that a Christian celebration on 1 January 2000 is wholly justified on the basis that it is honouring the two-thousandth year since 1 BC, being the holy year of Christ's nativity.

If there is any problem about which the whingers are justified in complaining, it is the naming of the year 1 BC. 'BC', after all, stands for 'Before Christ' (in Latin it is AC – *ante Christum*), and obviously there appears to be something a little bizarre about naming the year in which Christ was supposedly born '1 *Before Christ*'. But this was not the fault of Dionysius. He hadn't thought this bit through in 532 AD, for it was never part of his original canon, and, in any event, it must be remembered that his new dating system was not instantly taken up when it was first introduced in 532. It took several hundred years to catch on.

The mistake that was made *several hundred years later* was in not giving the actual year of Christ's birth a better name than 1 BC. It could have been called something like *Nativity Year* or *Christmas Year* or something tinselly like that. The year after *Christmas Year* would remain as is (AD 1) and the year before it should have been called (instead of 2 BC) 1 BC, and years would have carried on sequentially backwards from there. It is true that in terms of mathematics, this would have been far more convenient, for it is frustrating that there is no given year between 1 BC and AD 1. But to call it 'zero', as we have seen, would not be a satisfactory solution. We just have to remember, when making a calculation, that there is no interim year here. For instance, if you ask yourself in 4 BC what year it will be in eight years' time, the instant unthinking answer would be AD 4. But that, on reflection, would be wrong. Eight years on from 4 BC takes you not to AD 4 but to AD 5. Why? Because there is no year between 1 BC and AD 1. We can solve the problem in the short run; in the long run, though, we can only sit back and marvel at the querulousness of man: 'It is in disputes, as in armies,' says Jonathan Swift, 'where the weaker side sets up false lights, and makes a great noise, to make the enemy believe them more numerous and more strong than they really are.'

In this case the noisy pedants should be sunk without a trace.

CHAPTER XII

AETATES

Eras

The Egyptian input into the development of our strange modern calendar has already been well established. They were the first to move away from the moon and on to the sun as a means of calculating the length of a year. They worked the year out to be 365 days long, apportioning it as 12 months of 30 days each, with five special festival days thrown in for good measure. These last were designated the birthdays of Osiris (the many-eyed mummy-ox), his wife and sister Isis (vulture head), Horus (falcon face), Nepthys (the horned one) and Seth (a composite creature with a greyhound's body, slanting eyes, square-tipped ears and a long forked tail). It was Isis who apparently said: 'I am that which is, has been and shall be. My veil no one has lifted. The fruit I bore was the sun.'

It is fortunate that we did not inherit the irksome need to celebrate the birthdays of Falcon-Face and his Egyptian friends. Even so, we did take the 365-day year and the 24-hour day from Egypt. Another important Egyptian innovation was the giving of dates to years. They did this by numbering from the first year of each pharaonic reign. What we now call 1360 BC would have been known to the Egyptians as the second year of Tutankhamun; likewise 1200 BC was called the fourth year of Ramses II. This system was used right through until the sixteenth century in many places around the world. In England it has always been used on important parliamentary documents. For example, 6 Henrici. III means the sixth year of Henry III's reign (i.e. AD 1222).

In the twentieth century the term 'era' has been appropriated by geologists wishing to make sense of the immense age of the rocks and minerals they examine, each of which dates from a time before there was any official structure to human chronology. A larger unit than the era, the aeon, or eon, is also used by geologists to group eras together, though in religion the aeon specifies an age and can be used as a synonym for a god or any other eternal being. Basilides, that eccentric Egyptian Gnostic (who claimed to have obtained secret, esoteric knowledge from Glaucias, an interpreter of the Apostle Peter), was convinced that universal time was constructed of 365 such aeons, but his great rival, Valentinus, reputedly the author of the Gospel of Truth who also lived in Alexandria in the second century AD, claimed to *know* that there were only ever 30 aeons. These debates have lost their edge over the years. We are nowadays interested in aeons or eras only if they help us to put historical events into some form of chronological order.

The three main eras of geology are the Palaeozoic, the Mesozoic and the Cenozoic. To distinguish between them one is not, you will be relieved to hear, expected to know anything about sedimentary rock formations. The Palaeozoic era started approximately 570 million years ago and lasted until the beginning of the Mesozoic era, 345 million years later. This period is characterised by great shifts in the earth's crust, when large areas of North America were under warm, shallow water. Evidence of coral and primitive life forms, such as short worms and other invertebrates, have been preserved in fossils from the early part of this period. In the Upper Palaeozoic era which followed about 175 million years later in which animal and plant species thrived. There were sharks, of which the largest, a heavily armoured dinichthys, grew to nearly 23 feet (7 metres) in length, had a plated body and looked like a modern science-fiction machine. Also at this time a new range of reptiles started creeping about, and tiny, poisonous, airborne insects are known to have conquered the skies. The Upper Palaeozoic age ended in tears when, due to a global rise in temperature, 80 per cent of all amphibious, reptilian and cold-water species were extinguished. The northern continents were covered by vast swampy forests which were repeatedly washed over by seas, covered with layers of silt and drained, allowing the forests to regrow and the whole process to repeat itself.

The second era of geology, the Mesozoic era, was the thriving age of the dinosaur. These extraordinary creatures yawned and bellowed over most of the earth's surface for 160 million years. The metozoic archaeopterix, a remarkable fossil, proved to the world that birds were descended from reptilian dinosaurs, while the chondrosteiformes fish were probably the ugliest creatures that ever lived. We divide the Mesozoic era into three smaller epochs which will all be familiar to the schoolboy dinosaur enthusiast. They are the Triassic, Jurassic and Cretaceous phases. The oddest mystery of palaeontology is how and why the dinosaurs all disappeared, but disappear they did around 65 million years ago. The most likely theory is that a meteorite collided with the earth causing ubiquitous clouds of dust and dramatic changes in weather conditions, but it remains an unproven explanation.

The Cenozoic era (our current age), the last of the three great geological eras, is known as the epoch of mammals and flowering plants. It started as a time of continuing volcanic explosions, tectonic shifts in the earth's surface and the building period of all our mountain ranges: the Andes, the Himalayas, the Alps and the Caucasus. The rich spawning of mammal species led, in the last quarter, to the arrival of the tool-using *Homo erectus* in Europe and Asia. *Homo sapiens*, who was clever enough to seek warmth and shelter in caves, followed hot on *erectus*'s heels. The last two periods of the Cenozoic era are known as the Pleistocene (nothing to do with the nearly eponymous children's modelling material), which began two and a half million years ago, and finally the Holocene period, which covers the last 10,000 years, being the period of human civilisation and world dominance.

Before these three great and famous eras came the Proterozoic and Archaeozoic eras. The oldest is the latter, which extends, in effect, from the earliest years of the planet as we know it 4.5 to 1.4 billion years ago. This period was distinguished by the formation of the seas and the beginnings of life as a proliferation of single-celled organisms. In the following, Proterozoic age, sandstone, iron-beds, glaciers, shale and copper all came into being.

The people of ancient civilisations had never heard of the Cenozoic era, or the Mesozoic or the Archaeozoic ages before that. To ancient man (that is to say the Babylonians, Romans, Greeks and suchlike peoples) the world began a few thousand years BC. The Greek Church, the Russian Orthodox Church and the Freemasons were among many sects that used a dating system which started from what, in each of their opinions, was the year of the Creation. Their estimates as to when the Creation happened ranged from 3761 BC (the Jewish *Anno mundi*) to 5508 BC (known to the Orthodox Church as the Era of Constantinople). The Greeks at one time dated their years by counting forwards from the first Olympic Games of 776 BC. Dates like 102.3 O.E. are commonly found on monuments and scrolls from ancient Greece. In this instance 102.3 O.E. means the third year after the one hundred and second Olympic Era (games). Since the Olympics (then as now) were held every four years, 102.3 is what we now call 365 BC. This, like most dating systems, is retrospective. The first games to be recorded at Olympia were a modest one-day affair in which a few sprinters ran the length of the stadium. Nobody would have created a dating system from those Olympic Games if they had continued in this fashion, but, as it happened, after a few hundred years discus, javelin, long-jump, wrestling, pentathlon and chariot-racing were added to the Olympic disciplines. Then, as now, the top athletes became celebrities. Without television or newspapers these ancient athletes were praised by the poets and composers of their age who wrote songs and odes to their strength while sculptors translated their rippling muscles into marble.

As to the Romans, they started their own dating system from the time they believed to be the true date of the founding of the city of Rome by Romulus in 753 BC. These dates were known as A.U.C. (*ab urbe condita*, or 'from the foundation of the city'). The year 1000 A.U.C. would therefore be our equivalent of AD 248. In point of fact it is known that a settlement on the site predates their estimate by about 1,000 years, but that is of no concern. An alternative Roman dating plan called the Era of the Caesars or Spanish Era is reckoned from 1 January 38 BC, being the year following the conquest of Spain by Augustus. This system was popularly used in Spain, Africa and parts of the south of France, but by a synod held in 1180 its use was abolished in all the churches dependent on Barcelona. Pedro IV of Aragon (known to his subjects as El Cruel or El del Puñal – 'He of the Dagger') abolished the use of it in his dominions in 1350. John of Castile (the son of El Cruel's much-hated rival, Pedro of Castile) did the same in his own territories in 1383. In Portugal the Spanish Era was used right up until 1422. The months and days of this era are identical to the Julian calendar. To convert dates from the Spanish Era calendar into our own, all you need to do is subtract 38 from the year, unless it is a date before the Christian era (BC), in which case subtract 39.

In south-east Asia the era starts from the death of the Buddha, which was said to have happened in 483 BC. The Buddha's real name was Gautama Siddhartha (the Buddha being his title, a name that means 'The Enlightened One'). Gautama was a rich prince who, as a young man, lived a life of complete idleness and luxury, mollycoddled to the point of abstraction:

> I was delicately nurtured, exceedingly delicately nurtured, delicately nurtured beyond measure. In my father's residence lotus ponds were made: one of blue lotuses, one of red and another of white lotuses, just for my sake ... Of Kasi cloth was my turban made; of Kasi my jacket, my tunic and my cloak. I had three palaces, one for winter, one for summer and one for the rainy season. In the rainy seasons during the four months of the rains, I was entertained only by female musicians. I did not come down from the palace.

At the delicate age of 29 he went out for a drive in his chariot and noticed 'a sick man, suffering and very ill, fallen and weltering in his own excreta'. This and other similar sightings led Gautama to reject his life of luxury and set out into the world as an ascetic in search of truth. Away from his palaces he withheld from food and drink for such a long time that he came to look, by his own account, quite revolting. 'Because of so little nourishment, all my limbs became like some withered creepers with knotted joints; my buttocks like a buffalo's hoof; my backbone protruding like a string of balls; my ribs like rafters of a dilapidated shed.' Enlightenment followed, and so too did a number of disciples. By the end he was a greatly revered teacher. It is reputed that he chose to die, knowing that the right time for his death had arrived. He ate a plate of *sukara-muddava*, which some believe to have been pork, others bamboo shoots; others say it was mushrooms, and according to a vociferous minority it was a form of rice-pudding. In any event, whatever it was, the stuff gave him a terrible tummyache. On his deathbed he was visited by his disciples, and with his last words – 'Transient are all things. Try to accomplish your aim with diligence' – he died.

After his death the Buddha's following grew and grew. Millions of people united to praise him and honour his teachings, spreading the word right across south-east Asia to the peoples of India, China, Thailand, Japan and many other countries. The Buddhist era was instituted by the third Buddhist council, which took place in Patna in 247 BC.

Another notable religious era dates itself from AD 622 and affects the whole of the Islamic world. Muhammad ibn-'Abd-Allah ibn-'Abd-al-Muttalib, better known to us now as Muhammad the Prophet (AD 570–632) was a far rougher cove than the Buddha. He spent the last 10 years of his life waging 'holy wars' against the Meccans, whom he resented for not having believed that he, Muhammad, was the medium through whom God was communicating to the world. As the son of Abdallah, a poor merchant from a powerful tribe called the Quaraysh, Muhammad received his first revelation (a vision of God) on

Mount Hira at the age of 40. From then on his revelations and visions continued unabated; they were messages from God (Allah) and the angel Jibra'el (Gabriel), but still not many people were paying him much attention. In spite of the relatively few converts he had attracted, the guardians of the holy shrine at Mecca (members of Muhammad's own tribe, the Quaraysh) were so worried that Muhammad might undermine the prominence of Mecca as a centre of pilgrimage that they proceeded to persecute him and his band of followers. In AD 622 Muhammad, fed up with persecutions, packed his things and emigrated to the town of Yethrib (Medina). It is this emigration from Mecca, known as the *hijrah* or *hejerah*, which represents the starting date of the Muslim era. Muhammad himself dated his own documents and proclamations according to the memorable events in his own life, but it was not until seven years after his death that 'Umar I, the second Muslim caliph, introduced what is now known as the *hijrah* era. Muslim dates are written with the prefix A.H. which, in Latin, stands for *Anno Hegirae* ('the year of the *hijrah*').' Umar accordingly started the new era on 16 July AD 622, which corresponded to the first day of the lunar month of *Murrham*.

In AD 1844 another Muslim prophet, Mirza Ali Muhammad of Shiraz, a merchant's son from Iran, declared himself *Bab* ('gateway') to the *imam*, (the perfect embodiment of Islamic faith). It is surprising that this did not immediately lead to his arrest as a heretic, but it did not. For, like Muhammad in his early days at Mecca, this new prophet was ignored by all but a passionate band of his own loyal supporters. Later on the *Bab* decided he was more than just a gateway. He wanted desperately to be the *imam* itself and promptly declared himself as such. Unsurprisingly, when this latter proclamation came to the attention of the strict Islamic authorities, Mirza Ali Muhammad became as unpopular with them in Iran as Muhammad the Prophet had been to the Quaraysh officials at Mecca. In next to no time the *Bab* was arrested, put in chains and marched before a firing squad on 9 July 1850. It is said that the first volley cut the chains that were binding his hands and legs, leaving the prophet himself unscathed. This was used by his later followers as proof, if any were needed, that the *Bab* was indeed the *imam*. The second volley, however, put an end to him, and the *Bab*'s body was thrown unceremonially into a ditch. Several years later his corpse was retrieved by his disciples (members of the Bahai faith) and reinterred at a mausoleum on Mount Carmel in Palestine. The worldwide Bahai era was instituted from 1844, being the date on which Mirza Ali Muhammad declared himself *Bab*.

Our system of dating according to the birth of Christ (BC/AD) is, as we have seen, believed to be the handiwork of a sixth-century abbot called Dionysius Exiguus, known in English as Denis the Little, the Slight or the Small, depending on preference. His nickname, it is thought, was not earned because he had a physically diminutive frame but is supposedly in honour of the great man's modest, self-effacing nature. Dionysius has been mocked and despised by many historians because he wrongly calculated the birth of Christ. As has been shown, Josephus and Clement of Alexandria made it clear that Jesus was born in the reign of Herod the Great, which means he must have been alive in 4 BC (749 A.U.C.) or even a little earlier, but Dionysius fixed the nativity on 25 December 753 A.U.C. He had read a text (that still exists today) by Clement of Alexandria which, it has to be admitted, was not crystal clear in its explanation as to when exactly the start of Augustus's reign was supposed to have taken place. Dionysius's mistake was one that anyone might have made.

Despite his unfortunate error, there is no doubt that Dionysius Exiguus was in other ways a clever, remarkable and important man. He had a formidable intellect, he built himself a reputation as one of the foremost theologians of his age, he was an accomplished astronomer and a brilliant mathematician as well. Dionysius Exiguus lived such a long time ago that we scrape barrels to find anything about him. According to the sixth-century historian, Cassiodorus, he probably hailed from Scythia. We know he was summoned to Rome by Pope St Gelasius I, but he arrived there in AD 496 a short time after

Gelasius's death. Gelasius's successor, St Anastasius II (known for his congratulations to the Frankish Barbarian, Clovis I, on converting to Roman Catholicism), welcomed Dionysius and set him to work as a scholar, mathematician and chief archivist to the vast papal library.

By AD 525 (or, more precisely, by 1277 A.U.C. as the date was then labelled) the tolerant and benign Pope St John I had decided that the dating system should not be based on the foundation of the city of Rome; for there must be more important things than that to consider as a basis for counting years – Jesus's birth was the event that sprang most readily to mind.

As a pope, John I was noted for his brave attempts to unite warring nations and to bring peace to the bickering world. From 523 to 526 he succeeded in ending the Acacian Schism, thus reuniting the Eastern and Western Churches and restoring peace between the papacy and the Byzantine emperor, Justin I. This peaceful act happened to enrage a Christian Ostrogothic king called Theodoric the Great, who loathed Justin and mistrusted the accord he had made with the Pope. Justin accordingly passed an edict banning Theodoric's sect and encouraging his own people to persecute Theodoric's Christian subjects. Theodoric ran to the Pope in a fury, demanding that he remonstrate with Justin on behalf of his Ostrogothic subjects. The Pope's historic mission to Constantinople to persuade Justin to withdraw his edict against Theodoric was a failure. Incensed at Pope John's weakness, Theodoric (once his close friend) captured John and left him in a sealed chamber to starve to death.

Before he died, however, Pope John, and his trusty abbot Dionysius Exiguus, had in 532 reorganised the dating system. From that point forth, Jesus Christ's first year of life, (754 A.U.C.) was officially renamed by papal bull AD 1 (*anno Domini* 1, the First Year of Our Lord); and the year in which this was done consequently became known as AD 532. After Dionysius had completed his work on the dating of years, he is reputed to have set about making a collection of 401 ecclesiastical canons which included the apostolic canons and the decrees of the

Council of Nicaea (setting the date for Easter and producing the famous Nicaean Creed) and the councils of Constantinople, Chalcedon and Sardis. He is also known to have catalogued the papal decrees of St Siricius and Anastasius II as well as completing translations, from the Greek, of the lives of St Pachomius, the founder of communal Christian monasticism, and Proclus, the last important Greek philosopher.

All that is forgotten now. In fact, his Greek translations have been lost, but Dionysius is still remembered, above all, as the inventor of the BC/AD system.

In these our modern times, sensitivity towards the feelings of minority groups is expected of us all, and certain voices have been raised in remonstrance at the suitability of a worldwide dating system based on the birth of Jesus Christ. There is a solution for anyone who is anxious not to offend, or is keen to show his oneness with the modern way of looking at things. B.C.E. is used to stand for 'before the common era', just as A.C.E. can likewise be used to mean 'after the common era'. 'Common' in this case means 'united' or 'general', as in the Book of Common Prayer, and has nothing to do with social class. Ironically, B.C.E. means the same as B.C., and A.C.E. is the same as A.D.

Scientists, especially when referring to dates far, far in the past, are often attracted to the dating B.P., which stands for 'before the present' and bears no relation to the eponymous petrochemical company. For small, written measurements of time the B.P. system is absolutely useless. The date 1 B.P., for instance, means nothing to anybody unless they can tell when the present is. Which is all well and good when you are listening to a live speaker, but what happens when the B.P. date is written down? You will not be able to understand what date is meant unless you know the date in which the text you are reading was written. However, in discussing the Palaeozoic, the Mesozoic and the Cenozoic eras B.P. dating is clearly very useful. To talk about the Archaeozoic era as having started 4.5 billion years Before Christ does sound petty; 4.5 billion B.P. is somehow less absurd, and with vast numbers like 4.5 billion a B.P. date will still make sense even if your book is being read 6,000 years after you wrote it. But on this scale it all seems mind-bogglingly far away. These huge numbers are out of the realm of our brains to comprehend. Nor can we make sense of such elongated units of time, except by compressing them into simple definitions of 'before' or 'after'. The Cenozoic era came after the Proterozoic era. That makes sense even if the figures themselves do not.

Chapter XIII

AETERNITAS
Eternity

There are those who fervently believe that nothing exists which is eternal. To their way of thinking, eternity and all forms of a supposedly infinite system (i.e. infinite time, infinite space, infinite mass, gravity, speed or number) are hidebound, illogical and therefore do not exist.

'Don't be silly,' blurts the quick mouth with a slow mind. 'Of course infinity exists. How long is a journey round the circumference of a circle? How high can you count without adding one to go even higher? What happens when you divide 10 by three? You get 3.333333333333333 . . . and the threes go on for ever. Don't they?'

Not necessarily. We will deal with numbers right away.

First of all, we have to ask ourselves what numbers actually are, something we rarely, if ever, bother to do. We readily accept that two plus two equals four, but do we really understand what that means? Purists like to argue that a number (let's say two) is independent of any meaning outside of itself. Two plus two equals four is a statement of fact, which follows a mathematical principle which we have all learned at school and now accept as an irreversible tenet of truth. But can we really say that two plus two equals four is a true statement if we do not really know what 'two' means? Take an example: obviously, two dogs plus two more dogs equals four dogs. We cannot argue with that. We can go down to Battersea Dogs' Home and test it out empirically for ourselves. 'Fido and Jock, to heel! . . . Lassie and Hedwige, over here. Yes, that's four dogs. Hooray! Euclid's principles work.' But if we throw away the dogs for a moment and just say in an abstract sort of way 'two plus two equals four', how do we know what the twos are supposed to represent? And, if we do not even know this, how can we be sure what the answer is going to be? In other words, how can we be certain that, in every incidence, two plus two will always equal four?

For instance, two eggs plus two dozen eggs does not appear to equal four anything. It equals either two unequally sized batches of eggs or 26 eggs. We have to be careful with numbers. A number on its own (take seven) has no meaning unless it is related to things outside of itself, i.e. seven dogs, seven cats, seven dots or whatever. Schoolchildren and mathematicians use numbers to add and subtract all the time without questioning their meaning; it is a form of shorthand convenience. For the rest of us we use numbers only with specific meanings attached to them (eight dollars, four biscuits, etc.). We left abstract mathematics way behind in the classroom.

No one is suggesting a regime of correction over this issue. In the normal run of life, if somebody says two plus two equals four, you will not win friends or influence people by remonstrating, but when we are talking about infinite series of numbers this unique line of pedantry needs to be taken out of its box and dusted.

Philosophers sometimes call numbers *qualia*. It is a jargon word meaning that numbers are essentially like adjectives – they describe things in the same way as colours do. 'Blue' is similarly an adjective which means nothing on its own. That is to say we cannot imagine 'blue' without envisaging a blue 'something' in our mind's eye – a blue car, a blue balloon or, at very least, a blue area on a blank screen. Likewise 'fat' or 'foolish' are qualifying words which need a 'something' for them to qualify. A pair of sheep in a field can be qualified by the adjectives 'fat', 'foolish' and/or 'two'.

Now, if we think of any gigantic number, in what ways can it be said to exist? Take a random invention like

9872348723987428374. It exists as a row of 19 separate symbols (some of which repeat themselves) written on a piece of paper. Together they make up a single symbol – a number – which, as we have shown, is an adjective, but as yet the adjective has not been applied to any particular group of things. In other words the total number of *things* that can be said to exist as a direct cause of this number is 19; 19 separate symbols which add up to no more than a so far unascribed adjective. So to say that an unlimited number of numbers exists in the world is untrue, since there cannot be more numbers in the universe than there are things. By creating the massive number above, all we have demonstrated is the existence of 19 symbols, an adjective printed on one particular piece of paper at a specific moment in time, but we have not, alas, *proved* that all the numbers greater or smaller than 9872348723987428374 exist. The only way to do that is to write the whole lot out or to count out that many things. In the same way you cannot prove the existence of unicorns just by drawing horses with horns on a piece of paper. All you will prove this way is the existence of pictures of horses with horns. If numbers are indeed the same as adjectives then they describe things and that is their only function. Green, for instance, cannot exist, unless there is at least one thing in the universe that is green. To say that green exists because it is the product of mixing yellow and blue does not, in itself, prove the existence of green.

We can extend this line of argument to say that the highest number in the universe, at any moment in time, cannot be higher than the total number of *things* in the universe. To take the total number of things in the universe and to add one to it will not prove anything.

In order, then, for us to *prove* that numbers can be infinite we will need two important ingredients: infinite time (eternity) and infinite space. And so we have to conclude once again that time is a factor in everything. If the *present* is all that exists (because the *past* is over and the *future* has not yet come), then the only things that can be said to exist are the things that exist *now*. As we shall see in a moment, the fact that 'now' is a permanently moving feast caused St Augustine a lifetime of heartache and mental turbulence.

If we go back to the example of dividing 10 by three, we can say that three goes 3.3333 recurring times into 10, but does the .3 recurring really go on for ever? Once again, we can answer that it does in mathematical theory only, but in reality, of course, it does not. That is to say, if there is a computer in this world doing the sum, it has not yet had time to be doing so for ever. So far, in mathematics, nothing has been proven to be infinite other than in an entirely theoretical way. It is surprising how many bright minds of the past have missed this point. Bertrand Russell, a brilliant mathematician and a famous, if not innovative, philosopher, found himself as bewildered as the great Leibniz when it came to infinite number. He could not accept the nature of numbers as we have described them here, nor did he ever take into account the limits that time sets on the possibility of infinite number. For example, he says:

Take the series of numbers from 1 onwards: how many are there? Clearly the number is not finite. Up to a thousand, there are a thousand numbers; up to a million, a million. Whatever finite number you mention, there are evidently more numbers than that, because from 1 up to the number in question there are just that many numbers, and then there are others that are greater. The number of finite whole numbers must, therefore, be an infinite number. But now comes a curious fact: the number of even numbers must be the same as the number of all whole numbers. Consider the two rows:

1, 2, 3, 4, 5, 6, 7, . . .
2, 4, 6, 8, 10, 12, 14, . . .

There is one entry in the lower row for every one in the top row; therefore the number of terms in the two rows must be the same, although the lower row consists of only half the terms in the top row.

What about space and time, Lord Russell?

Zeno, with his puffing athletes, shows us that you cannot keep dividing space and time for ever: eventually you come to a minimum unit of space-time which cannot be made any smaller than it is. But if space and time cannot be infinitely small, is there anything to stop them from being infinitely large? One of the most pathetic patters of ancient Greek philosophy runs as follows: 'A is much greater than B; B is much greater than C. Therefore B is both great and small – how can this be?' We have to watch the Greeks. '*Equo ne credite*,' as Virgil so memorably put it. '*Quidquid id est, timeo Danaos et dona ferentis*.' ('Do not trust the horse. Whatever it is, I fear the Greeks even when they bring gifts.') On the whole they were brilliantly clever thinkers, advanced for their time and justly regarded as the fathers of all modern philosophical thinking, but they were, in other ways, able to reduce logic to a form of such crass banality that the purity of their finest thoughts is often submerged by the quantity and inanity of their weaker conclusions. Plato, who was known to say some very silly things, was himself aware of the ludicrous limits to which the logic of Protagoras and the cult of Sophism could be stretched, as this satirical dialogue from *Euthydemus* makes clear:

'Tell me, do you have a dog?'
   'Yes and a vicious one too,' said Clessipus.
   'And has he any puppies?'
   'Yes, he has. They are just like himself, vicious as hell.'
   'And so the dog is their father?'
   'Yes,' he said, 'for I saw him on top of their mother with my own eyes.'
   'Is the dog not yours?'
   'He most certainly is.'
   'Then since the dog is yours, and he is also a father, the

dog must be your father, and you must therefore be the brother of these puppies . . . Tell me, do you beat this dog of yours?'

Clessipus laughed and said, 'Ho! ho! ho! Heavens, yes! Since I can't beat you!'

'Then do you beat your own father?' he asked.

'Ho! ho! ho! There would certainly be much more reason for me to beat yours,' he said, 'for taking into his head to beget such a clever son. Ho! ho! ho! But I suppose, Euthydemus, that your father has benefited greatly from this mighty wisdom of yours!'

This is a baleful way of conducting an argument, but the Greek Sophists found it irresistible. At its best it is interesting, but no more than that. Here is an example of how the very principles on which we base our understanding of the world can be temporarily stunned by the cunning machinations of the Sophist: 'Stale bread is better than nothing. Nothing is better than God. Therefore stale bread is better than God.' These sorts of arguments have often been used to promulgate the cases for, and against, eternity. So beware! For instance: 'If time goes on for ever and ever, then it is infinite in its extent. It is therefore infinitely large. In comparison to this infinitely large unit of universal time the microsecond is an infinitely small unit of time. So too is a millennium. If this is so then both the microsecond and the millennium are infinitely small: ergo the one is not bigger than the other or, if Zeno is right and things cannot be infinitely small, then neither microseconds nor millennia can truthfully be said to exist.'

Luckily the human race has grown up since these ancient days and that sort of reasoning went out of fashion with the toga. But infinity has remained such an uncomfortable subject that throughout mankind's effortful history of grappling to understand it, not much has emerged of any concrete value. Many philosophers put forward the view that it is hopeless trying to understand infinity – such an understanding will never be possible, they say, for we could only ever succeed if we had infinitely large brains with which to think about it. Anything that is finite, like the human brain, cannot hope to succeed in understanding the infinite. The Danish philosopher Søren Kierkegaard took this line of argument one step further in an assault on logic. If logic is just one small part of human experience, he ventured, then we cannot expect to understand the whole of human experience by the use of logic alone, for one part will never explain the whole.

When Einstein declared to the world that nothing could move faster than light, a sigh of relief went up, for infinity is to many such a frustrating concept that they will always be delighted to hear when it has been proved not to exist. Speed was another scaled quantity (like time, space and all the others), which, it was once thought, had the potential to be infinite. Now, thanks to Einstein, it was shown that nothing can exceed the speed of light: 186,000 miles per second.

One down, 1,000 more to go!

Once Einstein had established the limits of speed, other potentially infinite systems were soon to collapse. If heat could be characterised as an indication of the speed at which atoms are moving around, and if nothing can move faster than the speed of light, then heat, too, must have its limits. At the other extreme it has been shown that there also existed a coldest temperature below which it was not possible to descend. Known as 0°K (0 degrees Kelvin or -273.15 degrees Celsius), absolute zero is represented by the complete cessation of movement in the atom.

So temperature, too, like numbers and speed, has now been declared an infinity-free zone, and many other scales, which at one time were considered to have the potential for infinite measurement, have now fallen by the wayside. Only time itself – and space, of course – remain as obstinate brutes. Time and space are set apart from all the others in that they are the vital constituents of any potentially infinite system. For anything to be infinite it needs an eternity of time and space to prove itself so. We have seen how space and time are locked together as 'space-time' and shown that if eternity exists then so, too, must infinite space.

With this under our belts we can argue with the full confidence of our convictions that if space-time is ever proven not to be infinite then infinity itself in all its forms and manifestations must be jettisoned out of the window, declared a nonstarter, an aberration caused by a combination of the shortcomings of mathematics, the false doctrines of religion and the unquenchable zeal of the human imagination.

That point has nearly arrived. But the scientific community is still abandoned in the way it uses words. Astrophysicists talk about 'infinite density' and 'infinite gravity' even if they do not strictly believe in their existence themselves. What they really mean is mathematically infinite gravity and density. In other words the point where their sums go into a free-fall spin of recurring decimals. They are describing conditions at the centre of a black hole – a singularity – and it is therefore important that they should be able to draw the distinction between 'eternity', in which time goes on and on for ever, and 'infinite time', in which time is unbounded because it has come to a standstill.

If the universe were one huge block of space-time, then Einstein's relativity would allow for all times – past, present and future – to coexist. That we can see only one plane in the time dimension (the present) is neither here nor there. When we look at a three-dimensional object – a cupboard, for instance – we cannot see all of its sides at once; nor would it be sensible to infer from our one angle of vision that the other sides of the cupboard do not exist. The same is the case for time. We see only the present; we may have memories of the past, and we know nothing of the future. According to certain thinkers we cannot even guess the future from what we know about the present.

The urbane Scottish philosopher David Hume (1711–76) shook eighteenth-century society with a new, forcefully logical way of looking at the world. One of his most famous arguments runs a bit like this: 'So far, every time I have put my hand into a fire I have burned it. I have also heard from others that every time they put their hands into the fire, they burn them too. But I cannot draw the conclusion from this that the next time I put my hand into the fire I will necessarily burn it.'

Hume's argument might run counter to every notion of common sense, but it is nevertheless irrefutably logical. Just because I have burned my hand every time I have put it into the fire so far, I cannot extrapolate any concrete reasoning from this which will *prove* to me that I will definitely burn my hand if I do the same thing again. This is not the same as saying 'I will not burn my hand if I do it again'. All Hume is trying to point out is 'I have not conclusively proven, on this evidence alone, that I will burn my hand if I put it into the fire.'

Sir Karl Popper (1902–94) was not as urbane as Hume. In fact, by the end of his life he was known to be an irascible curmudgeon. Born in Vienna, he made his reputation as a fierce critic of astrology, metaphysics and Freudian psychoanalysis (which he called pseudo-sciences). His most famous book, *The Open Society and Its Enemies*, builds on a plausible view of Marxism only to smash the ideology to pieces in his later chapters. In *Die Logik der Forschung* (*The Logic of Scientific Discovery*) of 1934, Popper takes Hume's point a step further. He accepted the idea that a scientific premise can be verified only by obtaining the same outcome from an infinite number of substantiating observations. In other words, if I put my hand in the fire an infinite number of times and burn it on every occasion, I may say that my hand will always burn if I put it in the fire. Although, of course, I will never actually be able to say that because, in the infinite time it would take to carry out this infinite number of observations, I will never have time to announce my conclusions. My hand would be reduced to cinders long before then anyway. Where Popper went on from Hume was to say that things could be verified only by a process of 'falsification'. What he meant by this is that although you cannot say 'all ants have six legs' solely on the basis that every ant you have ever seen has six legs, you *can* say 'not all ants have six legs', if you have seen ants with six legs as well as ants with four or three legs. 'Not *all* ants have six legs' is a *true* statement if you can produce ants with varying numbers of legs, but 'all ants have six legs' can only, according to Popper, be a true statement if you have infinite time to prove it so, and if, during that infinite time, you never come across an ant with any number of legs but six.

What the space-time theory suggests is that all time is out there coexisting. But how can this be? When we look at a star we are looking backwards in time, we are seeing what it *was* like millions of years ago, not what it *is* like now. In the years that it has taken for the light from the star to reach us, all sorts of changes may have taken place. The star might have burned out, for instance. Whatever has happened, we cannot know without whizzing off towards the star ourselves at a ridiculously high speed. All we can say is that the 'here and now' for us is probably a very different thing from the 'there and now' for it. Any single atom which is deemed by us to exist in a certain place at a certain time is also existing at other times and at other places, if viewed from a different perspective. And according to the theory, those perspectives can be reached by flying around at the speed of light (which unfortunately we cannot do, for according to another law of relativity the faster we move the smaller and heavier we become); alternatively, we could enter ourselves into a zone of extreme gravitational force – but that's out too: our fragile frames would not allow it. If we can never achieve these perspectives and never get the opportunity to see another dimension of the space-time continuum, we can at least fantasise on the legacy that the Einstein-Minkowski space-time theory has left us.

If all times coexist, albeit on a different plane, our deaths, our lives, every decision we have, or ever will, take in the past, the present or the future, including the beginning and the end of the universe, they are all out there, somewhere inhabiting their own present. This is not an easy pill to swallow, somehow; we instinctively wish to reject it. As Einstein wrote to the recently widowed wife of his close friend, Michele Besso: 'The distinction between past, present and future is only an illusion, even if a stubborn one.'

Visionaries like William Blake and St Augustine have stood bravely where angels fear to tread. At the beginning of his spiritual quest, *Four Quartets*, T.S. Eliot contemplates the infinite present, asking whether the present and the past are not 'both perhaps present in time future, and time future contained in time past'.

In *Jerusalem*, another visionary religious work, Blake announces: 'I see the Past, Present and Future existing all at once.' It seems to be the way of the world that while the pragmatist does everything in his power to reject the theory of coexistent time, the man of religion has always striven to embrace it. What made Einstein's revelations so extraordinary was that they came, not from the mind of a messianic prophet, but from the mind of a scientist. St Augustine of Hippo, another who strove for and eventually witnessed visions of eternity, spent a great deal of his time ruminating on the evanescence of things and struggling within himself to define the meaning of time:

> What then is time? If no one asks me, I know; if I wish to explain to him who asks, I know not . . . How can the past and future be, when the past no longer is and the future is not yet? As for the present, if it were always present and never moved on to become the past, it would not be time but eternity.

Augustine's parents were peasant farmers, so it is said, tilling the fertile lands of the north African coast in AD 350. Augustine grew to be a man of intense intuitive belief. His mother, Monica, was a dedicated Christian, but his father, Patrick, was aloof, and their relationship together was joyless and foreboding. Even so, Patrick was a generous enough father to appreciate both the value of education and the precocious merits of his son's intellect; and so, at great personal sacrifice, he sent Augustine to the university at Carthage, where the young man effortlessly graduated in rhetoric. Augustine learned Latin, which he much enjoyed, Greek, which he loathed, and Hebrew, which helped him to read some of the original Bible texts. Soon he encountered a problem which was to dog him for the rest of his life. How come the Bible was so full of vile aggression and improbable rubbish and yet it was the word of God? It is this battle that identifies Augustine for us as a modern. He was very close to his mother, and was heard reciting Bishop Ambrose's delicate hymn *Deus, creator omnium* over 1,000 times out loud when she died. A suggestion has been made that he and his mother, Monica, may have enjoyed an improper relationship together. Five days before she died they were in Italy, heading back for Africa. At Ostia, the port of Rome, Augustine had a vision, which, in a moment of ecstatic bliss, he shared with his mother. It was a vision of *cara aeternitas* – 'beloved eternity' – which he later wrote up in his *Confessions*:

> Forgetting those things that are past, and reaching for those that are before all Time, by concentration of energy I press towards the garland of my heavenly vocation. As it was and so it will be always: indeed 'as it was' and 'so it will be' are not appropriate, but just 'is', because wisdom is eternal . . . And while we were thus talking of eternal life and panting for it, we touched it for a moment with a supreme effort of our heart. And we sighed, and left the first fruits of our spirit bound to it, and returned to the sounds of our mouth, where words begin and end. And what human word is like to your word, or like you our Lord, who remains himself for ever, without ageing and renews all things.

For St Augustine and his mother, there is no present, because the present becomes instantly past, and so, they would argue, we cannot in this temporal existence of ours experience a true present. St Augustine tells us that we know time only through the mind, which is constantly shifting from the immediacy of sight, which is all we know of the present (*contuitus*), towards the future (*expectatio*) or the past (*memoria*) with such rapidity that the soul is for ever distended (*distentus*) – stretching one way and then the other, in all directions without ever finding the truth. In order for us to understand the truth about time, we need to concentrate unbelievably hard (*intentio*), as Augustine claimed he was doing with his mother in Ostia.

There are echoes here of the Holy Trinity and maybe they are intentional, although Augustine never said as much. The three-in-one idea which manifests itself in the Christian belief as God the Father, God the Son and God the Holy Spirit is shown by Augustine to be reflected in the idea of time as a continuous present conceived (with the effort of *intentio*) as an eternal flux of *contuitus*, *expectatio* and *memoria*.

In one way the space-time continuum gives hope to eternalists, and in another it rudely shuts hope out. The tiresome upshot of space-time is that it rekindles all the old chestnut arguments about determinism. Do we have a free will? If the past, the present and the future all exist in this universe now, then what is the point of our ever doing anything, if in fact we have already done it, in another area of space-time? Have we really made any choices in our lives, or were they already made before we were even born? Schopenhauer had already decided that free will did not exist. He based his argument on causality. Everything that happens must have a cause, the cause is the thing that makes it happen. Without causes, nothing could happen. If there is a cause for every cause then we spiral away in an infinite regression of causes. Schopenhauer's implication is that we cannot apparently do anything as an act of our own free will, because the will itself must have been caused by something else and that 'something else' must have itself been caused by another 'something else'. If this is so, then according to Schopenhauer the free will cannot exist. Schopenhauer has been upbraided as a pessimist for his views, but in his own lifetime he strongly denied the charge. If we have no free will, he said, then we cannot be held responsible for our actions. We need not feel any guilt, any shame or any remorse. That makes life much easier.

This must always be the case with visionaries, like Blake and St Augustine, who are obsessed, striving throughout their lives for a momentary glimpse of eternity. Eternity for them did not mean time that goes on and on for ever – that would be boring in the

extreme. What Blake and Augustine claimed they had seen was
the whole of Minkowski's space-time block in one ecstatic vision.
And others claimed to have seen it too: Isaiah, Muhammad the
Prophet, and the Buddha who, it is said, was even able to glimpse
his previous incarnations. Henry Vaughan turned his own view
into a light, trite rhyme:

> I saw Eternity the other night,
> Like a great ring of pure and endless light,
> All calm, as it was bright,
> And round beneath it, Time in hours, days, years,
> Driv'n by the spheres
> Like a vast shadow moved; in which the world
> And all her train were hurled.

CHAPTER XIV

TEMPUS OBSOLETUM
Primitive Time

Can we imagine what it would be like if we did not know all the things that we have been taught? We were brought up in the understanding that the earth, which is spherical, orbits the sun once a year, and that the moon, which is also a sphere, orbits us once every 29½ days – with that speck of knowledge it is quite impossible for us to see things any other way. The temptation, when we look back at our distant ancestors, is either to ridicule them for their backwardness or, as most historians are inclined, to patronise them for their sophistication.

Time and again we hear it said that the builders of that most mysterious monument, Stonehenge, must have been gifted and intelligent people, way ahead of their time, wondrously sophisticated, dragging 20-tonne blue stones from the Prescelly Mountains of west Wales the distance of 240 miles (386 kilometres) and erecting them in an intelligible pattern on Salisbury Plain. A cynic might look at the Stonehenge builders in a different light. Times were undoubtedly hard for them 3,000 years ago; food was not farmed in the structured way that it is today. Brute force, torture and ritual murder were evidently part of life's tough routine in those days. Did they really need to add to their woes by dragging those colossal rocks all that distance? Was it really so necessary? There must have been other rocks nearer to Salisbury

Plain; if not, why didn't they erect their edifice in the Prescelly Mountains and save themselves the pain of heaving them halfway across Britain?

To praise these people for their effort, or to condemn them for their folly, would be unnecessary; to understand their motivation would be extraordinarily interesting. We know little about these long-dead Britons, though we can assume that, like all human civilisations before and since, they ordered themselves into groups: the captains, kings, generals or high priests bossed everybody else about while furthering the interests of their own close friends and family. Fear motivated them all, just as it motivates us today. If the chief wanted a big pile of stones brought over from Wales, then the people made sure that he got them delivered on time and assembled in the way that he ordained them to be assembled.

The greatest question that still lingers unanswered about Stonehenge is what motivated these people to put it there? Let us assume that the underlings of the situation were simply obeying orders – the likelihood of it being otherwise is too remote. But what motivated the givers of the order? Fear, as Alfred Adler never tired of telling people, is the principal motivating factor for all our actions. Oscar Wilde was more specific. 'The terror of society, which is the basis of morals, the terror of God, which is the secret of religion – these are the two things that govern us.' What would the chief be so afraid of that he needs to build Stonehenge? There are plenty of stone circles elsewhere: the region around Stonehenge alone has six other sites. Was it simply a competition for who could build the biggest and grandest megalithic monument? But even self-aggrandisement, Adler would have said, is a product of fear.

The standing stones of Stenness in the Orkneys are not dissimilar from those at Stonehenge, and the custom of erecting circular sites like these was prevalent in Britain, Ireland, Scandinavia and Gaul during the whole of the Neolithic and early Bronze Ages. The Gauls and Picts, even the Vikings, nearly a millennium after the final touches had been put to Stonehenge, were terrified that the sky would one day crash on their heads. In modern time Swedes and Norwegians have denied that the Vikings ever wore helmets with horns sticking out, but is it possible that the builders of Stonehenge were protecting themselves from the terrible day when the sky might collapse upon them?

We do not know and probably never will. While archaeologists have made considerable headway in understanding the different phases in which Stonehenge was erected, dismantled and rearranged, the most pertinent questions about the site remain unanswered. Many believe (are they wishful thinkers?) that the stones were part of a temple for sky-worshippers. It is certainly true that if you stand night after night at the 'Altar Stone' – that massive slab in the central circle – you will be able to see the various risings and settings of the sun, moon and stars aligned along certain axes with the surrounding stones. But are these alignments significant or merely coincidental? After all, anything could be used in this way – any circle of stones would work even if they were plonked down without the slightest regard to the heavens. The clock-maker John Harrison invented a chronometer that was accurate to within a second a month. To measure that degree of accuracy, Harrison sat up through the small hours, in his home, observing certain stars as they appeared to pass through the gap between his window frame and a neighbour's chimney. But the piece of evidence that most excites the mystics and their New Age companions about Stonehenge is an apparent alignment between the rising sun on midsummer day with the 'Heel Stone', a 35-tonne block of sarsen sandstone set in an avenue some 82 feet (25 metres) from the outer circular ditch. One does not wish to pour cold water over the sun-worshipping theory of Stonehenge – it may be true and, in any case, has provided centuries of enthusiasm to those wishing to explore the heavens through the gaps of these dark hewn stones, but as most serious historians will aver, the connection must be treated with caution.

If we stick with these early people for a while, or perhaps go back a bit to their Palaeolithic ancestors, we can only guess at how they might have viewed the world around them and how they best succeeded in measuring time. Some things we can at least be certain about. The sun, the moon, the stars and the seasons were all around when the earliest variety of *Homo sapiens*, with his low moronic brow, protruding jaw and pendulous underlip, loped his way out of Africa. One often wonders how we ourselves would have comprehended the world if we were just dumped in it without any prior knowledge.

Many ancient peoples believed that the sun and moon were of equal size. If you hold your thumb up to a full moon and try to do the same with the sun on the following day, you will see why they thought this; the two orbs appear to have exactly the same diameter. 'And God made two great lights: the greater light to rule the day, and the lesser light to rule the night.' Can we say, then, that these ancient people were foolish in that they did not consider the perspective of distance, that they had not realised the moon was 400 times closer and 400 times smaller than the sun? Even the earliest of earliest *Homo sapiens*, those fossilised things of the late middle Pleistocene epoch about 250,000 years ago, even *they* must have come with some baggage. Their parents – whoever they might have been – must have taught them something. So there cannot ever have been a time when the species *Homo* (or any other species for that matter) was relying for its knowledge of the world on pure conjecture and empiricism. There must have always been a certain amount of parental programming, system software if you like, entering the brain as the young *Homo* was growing up. We believe that the modern system software is infinitely superior. We are taught from the earliest age that the world is round, but Palaeolithic man was led to believe, by his ill-informed parents, that the world was flat. We are cleverer than ancient man because we are programmed from birth with more useful and more accurate knowledge, but that is not all. Geneticists agree (and it is hard to resist their view) that human intelligence has vastly improved with the passing of generations, irrespective of the increase in effective knowledge that the intervening years have allowed us to attain. What do we mean by this? Well, primarily that we are much cleverer than our distant ancestors were, despite the superior knowledge which we hold – in other words, we could outwit them even without it. It is often said that one's own great-great-grandparents would seem like foreigners to us if we could meet them on equal terms.

So Palaeolithic *Homo* was not having to discover everything about the world for himself, as we sometimes imagine. He, like ourselves, was relying on the received wisdom of his teachers and parents mixed with his own experience. Was it one person who first discovered that the seasons are cyclical, or was this wisdom gradually acquired? We shall never know. If we were never informed that the seasons came and went in strict rotation, how long would it take for it to dawn on us that nature over the course of a year was conforming to a predictable pattern of events?

Obviously, farming was out of the question before *Homo* had learned to predict the seasons. If he did not anticipate autumn and winter following summer, he might well have filled himself in 'the season of mellow fruitfulness' and then starved to death over the ensuing months. If the beginning of agriculture coincides with the human ability to predict the seasons, then we can assume that the beginning of the history of any form of time-reckoning was about 9,000 to 11,000 years ago – not very long when you look at it one way. It was at this time we know that the Natufians of Palestine were going about with sickles. They might have been fixed-abode farmers – we cannot be sure. So too might have been the people of Ali Kosh, that border region between modern-day Iran and Iraq, where it is thought that wild barley and emmer wheat were cultivated around 7000 BC. But the proof of this is still wanting; so too is the proof that fixed-abode farmers were growing peas, bottle gourds and water chestnuts at the Spirit Cave in northern Thailand some 11,000 years ago, or that beans were being planted in the Tehuacan Valley in Mexico 10,000 years before Christ. Once again we simply do not know.

However, we can be fairly certain that the builders of Stonehenge were not just passing through. Farms must have been commonplace between 4,000 and 6,000 years ago. Before that, some animals were domesticated. That is to say they followed or were whipped along by nomadic peoples wandering hither and thither in search of fresh vegetation. Before these nomads had appreciated the benefits of keeping their own cattle, pigs and dogs, it is thought that *Homo* was a savage hunter-gatherer – and at this point, one is safe in proclaiming, a complete alien to most of his modern-day descendants.

And so it is to the settled farmers in the first six millennia before Christ that we look for the origins of the calendar. With the exception of certain, highly civilised communities – the Babylonians, the Mayans and, of course, the Egyptians – most cultures were not, it is thought, numerate. There is no evidence that our friends from Stonehenge had the remotest interest in numbers. So without the use of numerical calendars, such as we have today, these ancient people had to rely on their own primitive observations of nature – more than we do now when planning to plant the dahlias in our own back gardens. We can perhaps get a hint as to how this was done by looking at some of the tribal methods for counting time which have been recorded during the last 3,000 years.

Early written accounts are few and far between. The most famous is by one of the earliest Greek poets, Hesiod, whose *Opera et Dies* ('Works and Days') gives a vivid picture of peasant life in the eighth century BC.

Hesiod himself was a shepherd, though it is thought that, by birth, he was nobler than this profession implies. His brother, we know, grabbed most of the family chattels when their parents died and subsequently bribed some judges from the local community to side with him against Hesiod when the issue came to court. One suspects that Hesiod's shepherding years were full of bitterness because of this, and he must have been overjoyed when muses appeared before him at Ascra, near Mount Helicon, and endowed him with poetic gifts, bidding him 'sing of the race of the blessed gods immortal'.

So Hesiod became a poet. In the second part of 'Works and Days' (thought by some to be the work of another poet) the seasons are described with vivid rhythmic beauty, and the author lays out certain primitive rules for effective farming. The time for ploughing and sowing is shown by the cry of migrating cranes, he says. If one sows too late, if one is deaf to the wailing cranes, Zeus alone may save the crop with the intervention of rain on the third day after the cuckoo has called for the first time in the leaves of the oak. He advises farmers to prune their vines 'before the appearance of the swallow' and to stop digging in the vineyards 'as soon as the snail climbs up the vines'. The time for killing goats and drinking the wine is the summer, evidenced by thistle blossom and the shrill pitch of the crickets – and here's an odd one: when the fig tree shows leaves that are as big as the footprints of the crow, then the sea can be safely navigated. This sort of rustic advice was still in use 25 centuries later.

Native Indians from Pennsylvania would say that when the leaf of the white oak reaches the dimensions of a mouse's ear then it is time to plant the maize. Did Indians walk up to oak trees with mice in their hands? Probably not. Hesiod, or whoever did in fact write the second half of *Opera et Dies*, had not invented the rules of which he spoke. They were part of the folklore of peasant wisdom at that time. Hesiod was simply observing them. Homer knew about the 'cranes fleeing winter' and so did that noble Dorian Theognis. 'I hear, son of Polypias, the voice of the shrill-crying crane, even her who to mortals comes as harbinger of the season for ploughing.' Watching the birds must have been an unreliable business; should anything go wrong, the birds catch a virus or get eaten by foxes, then the farmer fails to sow his seed; Zeus forgets to send the rain on the third day after the cuckoo has sounded and now a whole community is at risk of starving to death.

Aristophanes, the satirical playwright, was an Athenian citizen writing 400 years before Christ. He was wise to the folly of agrarian wurzels who aligned their livelihoods to the tweeting of birds:

All lessons of primary daily concern
You have learned from the Birds, and continue to learn.
They give you the warning of seasons returning.
When the Cranes are arranged and muster afloat
In the middle air, with a creaking note,
Steering away to the Libyan sands,
Then prudent farmers sow their lands.
The shepherd is warned by the Kite reappearing,
To muster his flock and be ready for shearing.
You strip your old cloak at the swallows' behest
In assurance of summer and purchase a vest.

from 'The Birds', tr. J. H. Frere

The Canadian Indians actually believed that the North American buff-collared nightjar, or whippoorwill, with its insistent shrill call (sometimes repeated 400 times a night without stopping), was actually nagging them to go and plant their seed. Modern farmers don't waste their time on this sort of thing any more, but it was not so long ago that they did. In 1821 a group of peasants from the Yorkshire village of Thorne attempted to wall in a cuckoo in order, they hoped, to enjoy an eternal spring. To this purpose they erected a wall around the bird, but the cuckoo was not impressed, flapped its wings and breezed over it. 'Ah!' retorted one of the peasants, 'another course would 'a' done it.'

'Turn your money when you hear the cuckoo,' goes one English saying, 'and you will have money in your purse 'til he come again.' This is patently daft, but we should not forget, going back to the earliest Bronze Age farmers, that nature to them was a stronger, older and wiser force than themselves and only a foolish man would ignore it.

In south-eastern Australia Bigambul bush tribes have carved up their year according to various tree blossoms. The *yerra* is a tree that blossoms in September and lends its name therefore to *yerrabinda* – the time of the year. Similarly, there are times of the year entitled *nigabinda* ('appleblossom time') and *wobinda*, at the end of January, which means 'ironbark tree blossom time'. Midsummer, when there is no blossom to speak of, is called 'the time when the ground burns the feet'. The Banyankole tribe of Uganda (now extinct but thriving until the late 1920s) believed implicitly that storms unfailingly signified the approach of summer and, as soon as they heard the first clap of thunder they would run outside dancing and shouting; they would tear their clothes into shreds and carry on dancing for many days.

Certain animal patterns have been observed to keep remarkable time, but even these should not be relied on as a source of time-reckoning. Timothy the tortoise, 156 years old at the time of writing, has been a pet at Powderham Castle, seat of the Earls of Devon, since 1845. For as long as family records can recall, Timothy (who in 1921 was discovered to be a female) has come out of hibernation and walked across the lawn on exactly the same day each year – or so we are told. European swallows seem to migrate at the end of August and return at the beginning of April regardless of the weather. Among different species of animals drinking from the same African water-hole, each has its own hour of the day for visiting and sticks to it with rigid decorum. Giraffes are observed to visit once every five days.

But neither these nor our own menstrual or circadian rhythms are good enough to make a practical calendar. Lawyers will confirm that in court any witness's assessment of time passing (was it five minutes or twenty minutes between the slamming door and the gunshot?) is flimsy and unreliable evidence. Pleasure and action make the hours seem short, or as Shakespeare would otherwise have it: 'Time goes upon crutches 'til love have its rites.' In any event the fact remains that man needed to and therefore did invent systems for measuring time, aware that nature's rhythms were not, in themselves, serving an adequate purpose. It has been argued that the necessity for a calendar runs deeper than a simple need to know when it is best to plant the cabbages. It is a psychological necessity. We know, for instance, that hostages and prisoners have made it their top priority to keep a count of passing days. In Defoe's classic novel *Robinson*

Crusoe, the eponymous hero, stuck on a desert island, expresses his fear that he might lose his 'reckoning of time for want of books and pen and ink and should even forget the Sabbath days'. The solution was to cut notches into a wooden pole immediately before breakfast every day. 'Every seventh notch was as long again as the rest, and every first day of the month as long as that long one; and thus I kept my calendar of weekly, monthly and yearly reckoning.'

Time is, and always has been, one of nature's great untamables, and the biggest problem is that we still do not understand what time is. 'Time is the King of men,' wrote Shakespeare; 'he is both their parent and he is their grave,' and so, like all untamable beasts, man's burning urge is to tame it. We see time as an old man with a long white beard and an hourglass in his hands; he is the inevitable, plodding, humourless grey face of destiny. For the ancient Greeks time was Kronos, son of Uranus and father of Zeus. To the Romans he was the sinister Saturn, an odious figure who devoured all his children – all, that is, except three indomitable boys: Jupiter, the god of air, Neptune, god of water, and Pluto, god of the Underworld. These three could not be destroyed by time. Saturn is the glowering god, symbolised by the gloom of lead and the venality of gold. For the Romans the Saturnalia was a festival of wildness when all restraint and decency were thrown from the window and the baleful influence of Saturn was celebrated in orgies of drunkenness, sexual excess and criminal misbehaviour. The courts were closed for the seven-day festival and crimes would, as a rule, go unpunished.

> Then rose the seed of Chaos and of Night
> To blot out order and extinguish light.
> Of dull and venal a new world to mould,
> And bring Saturnian days of lead and gold.
>
> Pope, *Dunciad*, IV, 13

Time is not, and never has been, a popular figure. He was, to Wordsworth, 'a galling yoke'. He is a destroyer, a thief, an unstoppable plod who turns our rosy cheeks to wrinkles, our heads to white and our brittle bones to dust.

> In the Burrows of the nightmare,
>   Where Justice naked is,
> Time watches from the shadow,
>   And coughs when you would kiss.

'O let not Time deceive you, you cannot conquer Time . . . Time will have his fancy, tomorrow or today.' These were the words of Auden, a man who towards the end of his days had the most wrinkled face in Christendom. Time has stolen him, too, just as it has borne our sons away and pilfered Oliver Wendell Holmes's hard-earned guineas:

> Old Time, in whose banks we deposit our notes,
> Is a miser who always wants guineas for groats;
> He keeps all his customers still in arrears
> By lending them minutes and charging them years.

It is no wonder that *Homo*, as he developed from *erectus* to *sapiens*, from savage brute to hunter-gatherer, from agrarian farmer to civilised man, saw more and more the need to control, to map, to order and above all to understand the most powerful and exquisite force of his existence. Those Stonehenge builders may have been boorish, even moronic by today's honed and better-mannered standards, but they have achieved something; for Kronos, Saturn, or whatever we choose to call him, has still not managed to topple the blue-stone megaliths of Salisbury Plain, and maybe he never will.

CHAPTER XV

TEMPUS NON SIMPLEX
Complex Time

O f all things that we do not understand, time is the most
aggravating. Not knowing exactly how it works puts
the kibosh on all our earnest attempts to make sense
of the universe itself or, as Paul Davies has put it:

> Until we have a firm understanding of the flow of time, or
> incontrovertible evidence that it is indeed an illusion, then
> we will not know who we are, or what we are playing in
> the great cosmic drama.

So what are the problems that we have in understanding time,
and why does Paul Davies think that we cannot understand
ourselves until we have sorted these problems out? Firstly, and
perhaps most frustratingly of all, it appears that we have not yet
managed to define what the word 'now' exactly means. This
is an ancient problem, one that gave St Augustine of Hippo
severe headaches for most of his adult life. The philosopher
finds himself continually strutting about in circles. He asks, first
of all, a question: 'If the past is over, and the future has not yet
come, all that exists is now; so how long does now last?' What
the philosopher is unable to work out is whether the past and
the future already exist, in which case 'now' is simply moving
along, somehow agitating each pre-existing moment as it passes,

like the sun which casts its gleaming rays on whichever side of the world is turned towards it, or (and this is another possibility) does the future not exist at all, except as an illusion created by 'now', which is constantly inviting new things to happen? This is where the philosopher needs to step aside and hand the reins over to the scientist.

We shall come to Einstein and the physicists in a moment, but firstly we ought to ask ourselves a few simple questions. Why is it, for instance, that we all seem to perceive time differently, not only in different ways from each other but to ourselves as well. Even with a clock ticking away beside us, we know that 10 minutes can sometimes appear to pass in a flash and, at other times, they seem to amount to an interminable duration. Often this is tied to the simple human frailties of wishful and negative thinking. For instance, if we are waiting for an important telephone call, for the kettle to boil or for our toast to brown, the time can appear to be stressfully stretched.

On the other hand, if you have an urgent thing to do before you eat your toast – stamp and address a letter, for instance – you might easily find that however fast you perceive yourself to have accomplished the task, your toast will be burned by the time you turn to attend to it. Either way you may wish to smash your toaster. This might simply be accounted for by assuming that the occupied mind is not paying attention to the passing of time and therefore, unlike the unoccupied mind, time will appear to be moving faster. But tests have shown that this is not the case. If you take a person and shut him into a dark cave for five days and then ask him, when he emerges into the sunlight, for how many days he considers himself to have been entombed, he will invariably guess a number that is lower than five. We are feeble and incompetent when it comes to estimating time, which implies that the official time we have constructed for ourselves, based on hours, minutes, seconds, etc., has little to do with 'real' time, if such a thing exists. Kant saw time as a form of human intuition, and some scientists of the twentieth century are coming round to this view.

Another philosopher, more homespun than Kant, was Mr Martin Dooley. Dooley was the fictional creation of Finley Peter Dunne (1867–1936), a humorous Irish-American journalist who wrote for Chicago newspapers in the early decades of the twentieth century. Martin Dooley's satirical, often risqué, philosophical musings were written in an imitation Irish dialect and were widely disseminated throughout America:

> Scales and clocks ar-re not to be thrusted to decide anything that's worth deciding. Who tells time be a clock? Ivry hour is th' same to a clock an ivry hour is different to me. Wan long, wan short.

In an experiment involving the quick succession of flashing lights, Daniel Dennett, an eminent scientist of the mind, showed that the concept of time is a very individual matter. This echoes the thoughts of the American writer William Faulkner, who wrote that 'Time is a fluid condition which has no existence except in the momentary avatars of individual people.' In Dennett's experiment, if a subject is in a darkened room, and two equally sized lights are flashed on and off with great rapidity, the subject will assume that they are the same light moving from A to B and back again. If the lights are coloured differently he will continue to labour under the same misapprehension and will claim that he can actually point to the spot where the light is changing colour. So what is happening here?

We know that when we watch a film we are seeing 25 still frames whizzing past our eyes every second, but we cannot see the joins. Or can we? We don't think we can, but maybe we are deluded; maybe there is something inside our brains that is editing them out. If we set up a circle of lights and made them flash on and off in sequence, and ask, 'What do you see now?', the experimentee (usually a student trying to clear his debts) will think that he is seeing one light following a smooth trajectory.

At this point the scientist needs to investigate for himself the meaning of 'now' as it was interpreted by the student in his answer to the question. What can the scientist conclude? There must have been a 'now' when the student saw the first light; a 'now' when he saw the second light; and there must have been another all-embracing 'now' when the student's brain filled in the trajectory between the lights to give him the impression that it was the same light that had moved, and not, as was really the case, different, adjacent lights flashing on and off. So there are three 'nows', are there? To put it less babyishly, the student's 'now' must have lasted at least as long as it takes for two successive lights to flash and a little bit extra for the brain to go back and reconstruct a bogus trajectory between them.

Remember the Zeno paradox about an athlete being unable to reach his finishing post? That taught us that time (as well as space) cannot be divided into units smaller than the smallest limit. We called these units *minima*. If we decide to accept Zeno's premise (and not everybody does), we should make it clear that there is a distance of space and time which is so small that it cannot be meaningfully divided – that is all. Does this mean that all distances of both time and space are made up of these discrete *minima*? In theory, yes. For if a *minimum* is the smallest possible unit of space and time, then any other larger unit (the hour or the mile, for instance) should contain an establishable number of *minima* within it.

It has been suggested that Zeno's purpose in presenting these paradoxes was to argue the case of his master, Parmenides, who had arrived at a conclusion, following his own path of trusty sophistry, which stated that *Being* consisted of only one thing (i.e. everything that existed was one) and that this one thing was motionless. If this is really what Zeno was trying to say, then we can interpret his paradoxes as being intended to cause the utmost confusion and logical disarray. The world that Zeno and Parmenides are trying to promote is a world that consists in a 'single being' which cannot be divided into smaller parts; any attempt to do so would result in absurdity, as Zeno's paradoxes set out to prove. In reality we can see that his paradoxes are far more important than the specious use to which he himself applied them, and it may be the thought of Zeno's own conclusions to his paradoxes that has thrown them into disrepute. The following two paradoxes by Zeno are supposed to contradict the logical conclusions of the others, i.e. the theory that time and space cannot be infinitely divided:

1. The Paradox of the Moving Arrow states that when an arrow is flying through the air, at any specific instant in time, it occupies only the space which is exactly equal to its own shape and size. In this sense, at any precise instant in time it might be said that the arrow is not in motion: imagine it as a photograph, for if it were moving the moment in time would be subdivided. So the arrow's journey can be seen as a series of motionless instants. Now, if the distance which the arrow has to travel in order to arrive at the next instant is so small that it cannot be subdivided, then how can the arrow move at all? Or, to put it another way, if every moment of the arrow's flight can be captured as a motionless instant, then how can it be said to be moving?

2. In the Paradox of the Stadium, Zeno asks us to imagine two carriages (we will pretend they are train carriages, even though these did not exist in Zeno's day). One is motionless and the other is moving past it. A third carriage is also moving past the stationary one, but coming from the opposite direction to the first. The point Zeno tries to make is far from clear, but various interpretations have been put forward. Either he is saying that it is impossible for the two moving carriages to pass each other without drawing level at a point in the middle of the stationary carriage (which he thinks should be impossible if the carriages were the length of just one *minimum*), or he is trying to argue that it takes the same amount of time for the moving carriages to pass each other as it takes each of them to pass the stationary carriage.

Either position is silly, and it is a shame that such a formidable thinker as Zeno should have come up with such a badly thought out argument as this. The arrow paradox is likewise flawed, since the definition of a *minimum* is not 'a distance that is so small that it takes no time to cross it'. On the contrary, it is the smallest division of space which takes the smallest indivisible unit of time to traverse. There is space and time in a *minimum* – it just cannot be subdivided.

That Zeno failed to see the weaknesses of the arrow and the stadium paradoxes is a pity, but the bigger shame (at least for his posthumous reputation) is that he failed to exploit the strengths of the athlete paradox as a useful means for the non-scientist to get a better understanding of the world around him. It is this uncertain mix of brilliant vision and sudden blindness that makes so many of the Greek philosophers absorbing and yet so frustrating to read. It has been said that Zeno was a drunkard. The wine of Elea was known to be hopelessly intoxicating.

In the first chapter of this book we left our discussion of Zeno's *minima* with a terrible paradox hanging in the air. The paradox, if you remember, was far more tricky than the false paradoxes of the arrow and the stadium, and it grew out of the logical conclusions that we drew from the story of Zeno's athlete. It went like this: if there is a minimal unit of space, then to cross it must take the minimal amount of time (since such a unit of space cannot be divided and therefore it must take a minimal amount of time to cross it). If a beam of light travels the distance of one *minimum*, then it must take the same amount of time as a slug. If a mile is made up of x *minima*, then, by extension, the beam of light and the slug will complete the mile in the same amount of time. This is how we left it. And this is where the great Albert Einstein steps into the fray, to rescue us from what appears to be an absurdly irreconcilable paradox.

His theory is the theory of relativity, which means very much what it says it means. It is a theory about how things like time and space are perceived in relation to other things, like energy, gravity and motion.

Many excellent and clear-sighted books explain with ease and simplicity how Einstein developed his two main theories of relativity. It is not the intention to precis them here, but only to show what Einstein succeeded in proving and how it brought about a conceptual revolution in our understanding of time. So please take your hows and whys to other books.

Perhaps the most important discovery that Einstein made was that light moves at the fastest speed there is. Nothing can go faster than light. So, you may well ask, what has that got to do with time? Speed is, of course, a measurement of distance over time, as in miles per hour, metres per second, etc. In the case of light, the speed is 186,000 miles per second.

When two lorries, each moving at 40 miles an hour, drive towards one another, they are coming together at a speed of 80 miles an hour. In other words, the distance between them is shrinking at the rate of 80 miles an hour. So, what happens when two beams of light travel towards each other, just like the two 40-mile-per-hour lorries? If light moves at 186,000 miles per second, then the speed at which the two beams are coming together should be 186,000 plus 186,000, which equals 372,000 miles per second. But Einstein tells us that 186,000 miles per second is as fast as anything can ever go, so 372,000 miles per second is simply not possible. So what can be done? How can this problem resolve itself? Surely there must be a difference in speed between two beams of light whizzing towards one another, and, say, a slug moving towards a beam of light; yet both the first example (of two beams of light) and the second example (of the slug and one beam of light) add up to a combined speed which is greater than 186,000 miles per second. 'Help! Help! We're stuck!' Remember that speed is distance over time. If nothing goes faster than light, then Einstein realised that the only way to account for the two light-beams coming together is to leave your speed at 186,000 miles per second and adjust the other parts of your equation, namely time and distance. So, does this mean that the speed remains the same, but time and distance change? Yes: Einstein has *proved* that time and distance are relative to motion.

This, in turn, leads on to something very interesting. We will use Einstein's own example of a moving train. Let us assume that a person (we will make it somebody memorable, like a nun) is sitting exactly in the centre of an empty railway carriage as it hurtles along the track at high speed. Let us also assume that on the embankment, as the train rushes past, stands another very memorable person (how about a beekeeper?). Now, the nun is a curious lady who enjoys nothing more than pressing a green button which makes two bulbs simultaneously light up, one at the front of the carriage and one at the back; and so, for miles on end, she switches them on–off, on–off with a beatific smile on her face each time she sees the two bulbs lighting up, on command, to the pressure of her skinny hand on the green button. She takes a rest and places two travel-sickness pills on to the back of her tongue; then she wanders off to find a glass of water. Now what? 'Oh, yes, my lovely green button.' She totters back to the middle of the carriage and, as she presses the button, notices a beekeeper out of the window staring at the train. At this point we must allow the life stories of the nun and the beekeeper to unfold elsewhere. Let us just assume they got married and lived happily ever after. All we are interested in here is that split second when the nun pressed the button and she, in her position at the very centre of the carriage, passed the beekeeper.

There is no reason to suppose, as far as the nun's view of things is concerned, that that particular button-pressing moment was any different from any of the others. For the umpteenth time, she saw the two bulbs light up simultaneously as she pressed the button. She and the bulbs were all moving along together, with the carriage. When you throw up a ball and catch it in a moving train, you do not expect the ball to fly to the back of the carriage because of the speed of the train. Similarly, the nun's button, which lights the two bulbs at either end of her carriage, will not behave any differently, from the nun's point of view, if the carriage is whizzing along at 100 million miles per hour or if it is standing still at a station. The nun will, in each case, see the bulbs light up at exactly the same time.

But what about the beekeeper? Why bring him into it? He is a key figure who will help us to unlock the mystery of relativity. At that instant when the beekeeper is standing on the embankment, exactly parallel to the centre of the carriage as it hurtles past, does he also see the two lights come on simultaneously? If the carriage were stationary, the answer would be yes. But it is not; the carriage, as we have said, is hurtling past him at 100 million miles per hour. This would imply that the light signal from the back bulb should reach him before the light signal from the front bulb, because the back of the train is coming towards him at 100 million miles per hour, and the front of the train is moving away from him at 100 million miles per hour. The answer is that the beekeeper does see the light from the back bulb before he sees the light from the front bulb, but not for this reason. Remember, Einstein says that nothing moves faster than light. So the light coming from the bulb at the back of the train *cannot* travel to the beekeeper at the speed of light *plus* the speed of the train; it only goes at the speed of light and no faster. Once again, the imbalance in the equation (speed = distance over time) is settled, not by increasing the speed (for that cannot be done), but by shortening the time. In other words, time is perceived differently by the beekeeper on the embankment (who sees first the back bulb and then the front bulb come on), than it is by the nun in the train (who sees both lights come on simultaneously).

The question that immediately springs to mind is: Who is right? If it is only a matter of perception, then surely either the beekeeper or the nun must have got it wrong? Do the lights come on simultaneously, or don't they?

Einstein's answer is simple. It is not a question of perception; it is a physical fact that time and distance are relative to speed. The idea that distance should vary according to speed is a strange one, but no less true for that. Einstein *proved* that the beekeeper, should he choose to measure the length of the carriage from his position on the embankment as it whizzed by, would get a different measurement from the one the nun would get if she measured the carriage from inside the train.

Einstein's relativity opened up a whole world of fun ideas, particularly when the theory was expanded in a second paper by Einstein which involved gravity, energy and mass. Relativity shows that the faster you go, the shorter time becomes. If you could travel at the speed of light (which, sadly, another of Einstein's rules of relativity forbids), time and distance would shrink into practically nothing. Going from one star to another, a distance of 40 light years, would happen in an instant, as though you were just walking through a door from one room into the next, but an observer on earth would swear that you had taken 40 years to complete your journey. Both are right. Time is relative, like everything else in this strange world. The faster you move, the more slowly time moves. This has been proved beyond all possible doubt by an experiment using two synchronised atomic clocks: one was whizzed about in an aeroplane, while the other remained stationary on earth. When the two times were compared, it was shown that time had moved more slowly in the fast-moving aeroplane. It is only through understanding these strange phenomena that one is able at last to untangle the paradox of the slug and the light-beam. Zeno's space-time *minimum is* indivisible, but it is also relative to the speeds of the slug and the photon.

There were many new theories that popped up in the wake of Einstein's relativity. Suddenly, time travel became a real possibility; we were able to give dates to the distant stars in the galaxy, and to calculate when the big-bang beginning must have taken place. Einstein's achievement was monumental. Among the multitudinous historical prejudices that his theory overturned were the following:

- That time and space are absolute (this was Newton's firm belief).
- That things can accelerate their speed *ad infinitum*.
- That two events can happen instantaneously in all frames of reference.

- That light travels through space in straight lines (more on this below).
- That space is made up of invisible luminiferous ether (never mind what that means).
- That space can be thoroughly understood by Euclidean geometry (this too will be explained).

The point about things happening instantaneously was clearly made by the beekeeper and his future wife. But simultaneity is not just relevant to nuns and beekeepers and very fast trains. The only way we can measure time *at all* is by gauging when simultaneous events take place. To say 'I will have my lunch at one o'clock' means, if you wish to be pedantic and absurdly precise about it: 'When the little hand on my clock is pointing to the one and the big hand gets up to the 12, I will simultaneously shove a forkload of sausage into my mouth. Or, to put it another way, the delivery of a forkload of sausage into my mouth and the arrival of the big hand at the number 12 on my watch will be simultaneous events.'

'It ain't necessarily so,' says Einstein, quoting from *Porgy and Bess*, an opera he probably never saw, by George Gershwin. 'It all depends on the motion of the observer relative to the two supposedly simultaneous events.'

In 1915, 10 years after the publication of Einstein's *Special Theory of Relativity*, he submitted another paper to the Prussian Academy of Sciences. It was no less important than the first. Entitled *General Theory of Relativity*, it dealt not just with the effect of constant speed on time and space but also with the effects of acceleration and, most importantly, gravity. Einstein discovered that time and distance are relative measures that depend on gravity in just the same way as they depend on speed. Why so? In what might this connection between speed and gravity consist? When we pace about from A to B, we are used to that feeling of being grounded to the earth. The force of gravity, as we were all taught at school, pulls us, and all our belongings, towards the centre of the earth. It is for this reason, and this reason alone, that the upside-down Australians at the bottom of the globe are able to go about their daily business without blood rushing to their heads. We are all very used to the feeling of being pulled downwards – to us, it is perfectly normal.

Now try to imagine something distinctly abnormal. You are floating around in a lift (or an elevator, if you come from that part of the world) in the middle of outer space. There is no gravity; the lift is in the shape of a cube and uniformly painted white, so, as you float around within it, you have no idea which side is supposed to represent the floor, which is the ceiling and which are the four walls.

Suddenly, the lift begins to move and accelerates in any one direction. If this were an empty train carriage, we can easily imagine what would happen. If our nun were sitting in the middle of her carriage, nursing her lovely button, and the train suddenly accelerated violently, the nun would slide across the floor and smash into the back of the carriage. She is held to the floor by gravity and thrown by the force of acceleration to the back of the carriage. By contrast, in the lift, you have no gravity to worry about; you are simply thrown to the back of the lift by its acceleration. Einstein claims that, if the lift keeps going at a constant rate of acceleration, you will be able to stand up and it will feel exactly the same as if you were standing on earth. The side in the direction to which the lift is moving will become your ceiling; the opposite will be your floor. You will simply imagine you are standing up in a normal way. What he is saying is that the force of gravity, which pulls us downwards on earth, *feels* exactly the same as if it were a force of acceleration pushing us in a straight line through space. Acceleration is the same as speed. Therefore, speed and gravity must also be the same as each other in relation to their effect on time and on distance. Time and space are effectively stretched by gravity, which means that the bigger the planet you live on, the greater the gravitational pull to its centre, and the slower time will pass in relation to the smaller planets.

One of the most fascinating upshots of the theory of relativity is the theory of space-time. It is an idea which, in its simplest form, can be explained as the inseparability of space and time – they are both seen as dimensions of the same thing, whatever that 'thing' may be. To suggest that space and time are independent, so the theory goes, would be wrong. Already we have seen how space and time are linked by the speed of light, and, if you accept that all things are moving, how can it be possible to separate the two? Who better to get this particular ball rolling than Zeno, our old friend from Elea?

Zeno's final paradox is not unlike his first, but contains subtle differences which are worth investigating here. Remember the well-oiled athlete, with his fluffy hair blowing in the wind, as he tries to subdivide distances between himself and the winning post *ad infinitum*? Well, this time the athlete has a name. Zeno calls him Achilles. Let us assume that he is referring to the bisexual demi-god who was dipped into the River Styx by his mother, the sea-nymph Thetis. Thetis, not wishing to get her own hands wet, submerged the whole of her son's body into the water except the part by which she was holding him – his heel. The waters gave the greater part of Achilles' body invulnerability against attack, but he was eventually slain by Paris, whose arrow, guided by Apollo, hit Achilles on the heel and killed him. It was a pathetic end to a once mighty warrior. In his lifetime, Achilles was known for his handsomeness, ferocity in battle and agile limbs, all of which were depicted with great relish by an anonymous fifth-century Athenian vase potter, who is known to us today only as 'the Achilles painter'.

If Zeno ever saw the works of 'the Achilles painter' is not known, but one suspects that he had a collection of them stuffed away in some secret drawer in his bedroom. Perhaps that is why he chose Achilles as the subject of his second paradox.

Achilles' life was full, he was a busy man, but in Zeno's paradox he is caught in a rare moment of frivolous pursuit – racing against a tortoise. The problem that Zeno expounds to us is this. Achilles runs a race with a tortoise, determined that, even though the tortoise had beaten the hare in their last race together, he would not outrun the great Achilles as well. However, since it is obvious that Achilles can run faster than the tortoise, the race authorities, unmoved by the warrior's gnashing teeth and burning eyes, grant the tortoise a 10-metre head-start. They both set off at the blowing of a ram's horn. By the time Achilles reaches the 10-metre mark, from where the tortoise started its race, the animal has moved on to the 12-metre mark. 'Puff, puff!' Only two more metres and Achilles will overtake the tortoise. But no! When he arrives at the 12-metre mark the tortoise is no longer there; it has moved to a point further down the track. This continues to happen indefinitely, for however fast Achilles runs he must always get to where the tortoise was before he can reach the place where the tortoise is, and, since the tortoise is permanently on the move (albeit at a laughably slow pace), then, by the theory of it, Achilles should never be able to catch up with the tortoise, let alone overtake it and win the race.

Obviously, something is wrong here, for, as with Zeno's first paradox, in which we know the athlete can complete his race, we can also be sure in this instance that if Achilles is demonstrably faster than the tortoise, he will be able to catch up with it in the race – unless, of course, he falls asleep by the roadside, like the hare, or sustains some terrible injury to his heel. But in the event that neither of these problems occurs, we are right to search for the chink in the paradox's armour. We can always argue our way around the problem, but this does not necessarily solve it. We could say, for instance: 'Yes, but,' (because all refutations of Zeno begin with these words); 'Yes, but the way in which we describe Achilles' and the tortoise's motions demonstrates that, when Achilles is in the immediate vicinity of the tortoise, his speed is greater than that of the tortoise and he therefore overtakes it.'

This may be true, but does it disqualify Zeno's argument? Hardly. All it does is look at the problem through a different lens. Yet we are sure that Zeno, with his fixation on the decreasing distance between Achilles and the tortoise, must be missing

Table 3:

A graph showing how Achilles manages to overtake the tortoise

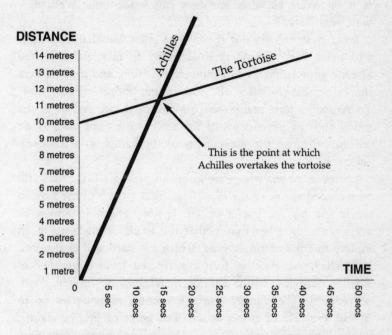

DISTANCE

Achilles

The Tortoise

14 metres
13 metres
12 metres
11 metres
10 metres
9 metres
8 metres
7 metres
6 metres
5 metres
4 metres
3 metres
2 metres
1 metre

This is the point at which
Achilles overtakes the tortoise

TIME

0   5 sec   10 secs   15 secs   20 secs   25 secs   30 secs   35 secs   40 secs   45 secs   50 secs

something. The paradox just cannot be right. So what is Zeno missing?

The question of speed is obviously crucial. Speed, as we never tire of pointing out, is calculated by dividing distance over time. If we draw the race as a simple graph it is easier to explain what is happening. In Table 3 we can see Achilles' thunderous black line as he pounds off at the rate of just over one metre per second. The tortoise has a 10-metre handicap, but they both start at the same time (shown by the bottom axis). If you follow Achilles' line you will notice that he has covered 11 metres in about six seconds at the point where he overtakes the tortoise. The tortoise, meanwhile, has travelled about one metre in 10 seconds. So although Achilles and the tortoise were running in the same direction along the same track, in terms of distance over time they were travelling along different routes. So what is distance over time if it is not speed? Is it the same as what scientists call space-time? Well, in a way, yes.

Isaac Newton was not the first to point out that both time and space were dependent on motion. In fact, Aristotle had already gone so far as to announce that time and motion were the same thing – after all, you cannot move without time. To Aristotle, time manifested itself through the phases of the moon and the movements of the sun, in the same way as we see time through the movements of the hands on a mechanical clock.

Lucretius, about whom we know surprisingly little, may still be regarded as one of the most important writers of the Roman world for his six books of *De rerum natura*, in which he argues, among other things, that the world is made up of an infinite number of tiny atoms, which are hard and changeless, and which are each in turn constructed from a number of smaller particles, which cannot themselves be broken down or reduced. The universe, says Lucretius, is made up of an infinite expanse of empty space. The point of life, he claims, is for each individual to sway the balance away from pain and towards pleasure. This can be achieved only by the study of philosophy, which he perceives as the only means to drive away the fear of death, which is the cause of so much of man's misery. Lucretius is an extremely interesting writer who has perceptive things to say on many subjects. His inquisitive mind is continually searching for the causes of things. Considering he wrote *De rerum natura* over 2,000 years ago, it is remarkable how many of his thoughts still appear germane to us today. His view of time as motion may have been taken from Aristotle:

And in the same way time itself cannot exist,
for only from the flight of things we get a sense of time . . .
No man, we must confess, feels time itself.
But only knows of time from flight or rest of things.

Newton's mistake (and who can blame him?) was to assume that time and space were eternally fixed, unwavering dimensions. A mile is a mile and that is all there is to it. An hour is an hour and wherever you are, whatever you are doing, in whichever part of the universe, that rule holds true, or, to put it in the great man's own words: 'Absolute, true mathematical time, of itself, and from its own nature, flows equably without relation to anything external.' Newton studied the laws of motion and came to the conclusion that bodies moved through space according to certain predictable laws. These rules are still in force today: 'Every body continues in a state of rest, or of uniform motion in a right line, unless it is impelled to change that state by forces impressed upon it.' Secondly, Newton showed that 'the change of motion is proportional to the motive force impressed; and is made in the direction of the right line in which that force is impressed'. His third law concluded: 'To every action there is always opposed an equal reaction.' To this was added his most famous law of all, the law of gravity, which is popularly rumoured to have been triggered by the fall of an apple on to Newton's head. His law of gravity states that the force of gravity between any two bodies is inversely proportional to the square of the distance between them.

With these and a bundle of other mechanistic rules, Newton, it was thought, could calculate anything and everything, turning the once mystifying universe into one gigantic, predictable clock. He was unquestionably one of the most remarkable men who ever lived. His contributions to physics, theology, optics, mathematics and a host of other disciplines put him apart from other mere mortals. He may have been a suspicious and bad-tempered man, but he was nevertheless hugely admired by his contemporaries, as one of them, John Locke, remarked:

In an age which produces the incomparable Mr Newton, it is ambition enough to be employed as an under-labourer in cleaning the ground a little and removing some of the rubbish that lies there in the way to knowledge.

But Newton himself saw it differently:

I do not know what I may appear to the world, but to myself I seem to have been only like a boy playing on the sea-shore, and diverting myself in now and then finding a smoother pebble or a prettier shell than ordinary, whilst the great ocean of truth lay all undiscovered before me.

Most of the principles of Newton's greatest thesis, the *Principia* of 1687, still hold good today. Einstein's relativity did not by any means destroy the foundations of classical Newtonian physics; in many ways it built on them. Newton's two biggest blunders were alchemy (he poisoned himself trying to convert base metal into gold) and his belief in time and space as absolute entities. He did, however, concede that our understanding of space and time are dependent on motion, and in this, like Aristotle, Lucretius and others before them, he was clearly right.

Try to imagine time in a universe in which nothing moves. It does not make sense. Time exists because things happen to make it exist, and these things – or events, as we call them – are caused, in all cases, by things moving in relation to each other. The same sort of argument can be applied to our understanding of space. What exactly is space? Well, it is distance, a measure of dimension. Is it possible to have space without anything inside it or anything bounding it, just pure space? The logical answer to that question has to be a firm 'no'.

Of course, we can imagine an area of, say, two cubic miles, with nothing whatsoever inside it, but that is not the same as empty, unbounded space because, in this case, it is bounded by the concept of two cubic miles, so what we are really imagining is an empty box, and a box is a kind of thing. However hard we try, we will be defeated in our attempts to imagine space without anything in it or around it. But we should not be too depressed by this, for how can we ever expect to imagine empty, dimensionless space? What, after all, *is* space if it is not the distance between things?

Once we have established that as a premise (and it is not an easy one to refute) we can proceed to ask ourselves another question which may, on the face of it, seem deceptively babyish. Is it possible to measure any distance in space without time? Babyish it is, but let's answer it just the same. No, we cannot do anything without time, and therefore we cannot measure any distance. But let's take human beings out of the equation. Let us suppose that measuring (that most human of all foibles) were not an issue, what then? Can distance between objects still exist if there is no such thing as time? We have to be careful here. Philosophers love to ask themselves a similar question about existence: 'If all the knowledge that I have about this universe has come to me, by one means or another, through my senses (i.e. through what I have seen, through what I have heard, felt, smelled, tasted, deduced and remembered), then what exactly is there that exists outside of me?' This branch of philosophy, known as epistemology, is a vast and complicated subject on which millions of words have been expended. Great thinkers like Plato, Hume, Kant and Schopenhauer spent their lives trying to work out what really exists beyond that which is apprehended by our own sensory organs. None of them arrived at a satisfactorily positive conclusion. A lifetime's search for a soupspoon of gain. Each of them made headway towards our understanding of what we cannot ever know. So much for the *limits* of human knowledge; the true extent of what we *can* learn about existence remains, for the time being at least, shrouded in mystery. But we do not wish to get embroiled with epistemology here, or we shall sink into a fathomless mire of circular argument. We must return to the burning question that triggered our little epistemological diversion in the first place.

'Can distance between objects still exist if there is no such thing as time?' No, it cannot. Distance, as we have seen, is the measure of space between objects or points. We have already shown that we cannot measure distance without time. Distance itself would not be possible without time in which to measure it. If objects ceased to exist, so too would space, since matter is the only measure of distance. And, since we have shown that objects (in the plural) cannot exist outside of time (for they are all of them in a permanent state of motion relative to each other), then, by extension, time cannot possibly exist without space.

You may need to check this through again. There are many different ways of putting this argument, and many are more complicated than this, but ultimately they will all arrive at the same conclusion. Even if you have not grasped every stage in the argument, you can get the drift: namely, that matter (objects, or things), time, motion and space (distance) are inexorably linked to one another. They are all interdependent. You cannot take any one element out of the mix and still expect to have anything meaningful left after the operation.

In 1908, a year before he died, Hermann Minkowski, a German/Lithuanian mathematician, brother of the then-famous pathologist Oskar Minkowski (who found that by removing the pancreas from a dog the animal developed diabetes, which led to the discovery of insulin as an antidiabetic), walked into a lecture theatre in Cologne and made a now-famous speech:

> The views of space and time which I wish to lay before you have sprung from the soil of experimental physics, and therein lies their strength. They are radical. Henceforth space by itself and time by itself are doomed to fade into mere shadows, and only a kind of union of the two will preserve an independent reality.

As we have just seen, anyone could have come up with Minkowski's basic idea while splashing in the bath, using nothing to work on but his or her brains. Without the scientific research, without any need for dull-grey Göttingen laboratories with their microscopes, test tubes, Bunsen burners and sickly smells, the humble Zeno of Elea might have reached the same conclusions as Minkowski while pondering his paradox of Achilles and the tortoise on the warm lapping shores of the Mediterranean Sea 2,000 years earlier.

Joseph-Louis Lagrange (1736–1813), the French mathematician, whose original name, Giuseppe Luigi Lagrangia, reflected his family's long sojourn in Italy, had already formulated an idea of four-dimensional space-time before the French Revolution, and his conclusions were developed from an intense study of *maxima* and *minima*. It is surprising how difficult it is to get these radical ideas to stick, for while Lagrange was the most honoured and fêted mathematician of all time (he was given apartments in the Palais du Louvre and Napoleon made him a Sénateur and a Comte de l'Empire), his views on space-time failed to take a hold for another 120 years. It was the added mathematical backing that Minkowski and later Einstein brought to the issue of space-time that made it all so tangibly exciting to the physicists of the early twentieth century. After 3,000 years of philosophers and mathematicians coming to the conclusion that space and time were both essential organs of the same body, Minkowski, late in the day, lumbers along to steal all the glory.

In fairness to Minkowski, he did have quite a bit of his own to add to the debate. He was a keen mathematician, and mathematics were, until the conceptual revolution brought about by Kurt Gödel (in his famous Gödel's theorem of 1931), believed to be the *sine qua non* of scientific proof. The battle between philosophers and scientists has been raging ever since the two professions went their own separate ways at the end of the eighteenth century. Nowadays, even the staunchest philosopher is forced to admit that laboratory experiments, mathematical ingenuity and laborious case-study observation will inevitably produce a higher degree of necessary proof than the abstract clockwork of his own logic.

What really made the difference between Zeno and Minkowski cannot be found in the quality of their rhetoric. While the former lolled around imagining the white gnashing teeth and curly locks of Achilles, Minkowski had slogged away at reams of maths until he had arrived at something which would show, not just that the space-time theory offered a realistic explanation of the universe (philosophers had already arrived at that), but *how* space-time effectively worked.

Hermann Weyl, a German mathematician and contemporary of Einstein, tried in his 1949 book *The Philosophy of Mathematics and Natural Science* to explain Minkowski's point:

> Reality is a four-dimensional world in which space and time are linked together indissolubly. However deep the chasm that separates the intuitive nature of space from that of time in our experience, nothing of this qualitative difference enters into the objective world which physics endeavours to crystallise out of direct experience. It is a four-dimensional continuum, which is neither 'time' nor 'space'.

What Minkowski managed to show was that time and space were not just interdependent dimensions of the universe, but that they were also, in a mathematical sense, geometrically equivalent, and related to each other by the speed of light. Some people find it very difficult to grasp the concept of space-time; others have no problem. In essence it describes a universe which is four-dimensional. We all know and understand the three dimensions of space (height, width and depth), but there is, according to Minkowski, another dimension of equal importance to the structure of what is out there – namely, time. These four dimensions are inextricably linked. They cannot be separated, or, as Einstein put it: 'It is neither the point in space nor the instant in time at which something happens that is real, but only the event itself.'

To make the whole idea more digestible, space-time can be explained by analogy to a sheet of latex rubber. Imagine that space-time is represented by a rubber sheet stretched across a frame, like a trampoline, and that on that surface parallel lines have been drawn running, as it were, north to south and east to west, forming a grid of evenly drawn squares. If we put a heavy ball on to the trampoline, what will happen? It will sink in the middle, causing the rubber sheet to curve, and the squares drawn around to be stretched and contorted by the weight of the ball. This far should not be hard to imagine. Suppose we rolled a small marble in a straight line across the surface of the trampoline. The trajectory of the marble will follow a natural curve and bend as it follows the curve around the heavy ball. So what does all this have to do with space-time?

Einstein and Minkowski demonstrated successfully that space-time works in exactly the same way. Suppose the heavy ball were a large planet like the earth. Due to the force of the earth's gravity, space and time are curved around it in just the same way as the surface of the trampoline curves under the weight of a heavy ball. So what was all that passing marble business about? Anything that attempts to proceed in a straight line past a gravitational force like the earth will find that the straight line he thinks he is following is, in point of fact, curved by the gravity of the planet. Remember the grid squares we drew on the trampoline? These represent two-dimensional units of space-time. As the marble passes the heavy ball, its course will curve, just as the lines on the rubber sheet do. So, as a body passes near a gravitational force, both time and space are curved in exactly the same way.

What is most interesting about this is the way it makes us think about straight lines. Euclid (c 300 BC), the father of mathematics, compiled the seminal work *Elements*, whose 13 books covered the mathematical principles which formed the basis of mathematics textbook teaching right through to the twentieth century. *Elements* contained all the standard theories and definitions, from Pythagoras's theorem to the angles of a triangle and the definition of a straight line. Space-time involves non-Euclidian geometry. To Euclid, a straight line is the shortest possible route between two points. (Tennessee Williams held that 'Time is the longest distance between two places'.) Euclid's definition of a straight line is the same for Einstein and Minkowski, though there is a subtle difference. For these scientists the shortest possible route between two points is not necessarily straight; in fact, parallel lines might, in extreme cases, be seen to converge. Remember how the grid is stretched and contorted on the trampoline? Euclidians might give a stubborn answer to all this:

'Yes, but [here we go again] if a line in Einstein's geodesic geometry is curved, then it cannot also be straight. It is either straight or it is curved, but it cannot be both. Surely it is a question of semantics whether we call it a straight line or not?'

'Ah ha,' comes the retort, 'but if the line follows the shortest possible route between two points, then it adheres to Euclid's definition of a straight line, doesn't it? Hence it is a straight line in Minkowski's space-time.'

'Oh, I suppose so. But can't we call it a geodesic straight line to differentiate it from a Euclidian straight line?'

'No!'

When we look at the geometry of three-dimensional spheres and curves, we notice that quite a few of Euclid's best theorems (which were plotted on flat sheets of paper) go awry. For instance, it has been held for 2,500 years that the angles in any triangle add up to 180 degrees. But what happens if you draw the triangle on the surface of a sphere? If you take a triangle, for instance, between three points on the earth's surface (New York, Moscow and Buenos Aires), the angles of the triangle will add up to slightly more than 180 degrees. That is just about easy enough to understand, but the trouble with four-dimensional space-time is that it is just about impossible to visualise, since anything we can imagine has to be in three-dimensional space. What we have with space-time is a block in which all space and all time are always present.

Newton's laws of motion had turned the universe into a predictable machine. Every turn of every planet in every orbit could be predicted by his mathematics long in advance of its happening. Time itself was predictable and dependable. God was a divine clockmaker who had created his toy, wound it up and lain back to watch it working unattended. If philosophers were concerned about a theory that set the future of the universe by a predetermined mechanism, the theories of space-time were even more extraordinary.

In 1974 the scientist Frank Tipler from the University of Maryland, pursuing the logic of Einstein's relativity to its bitter end, published in the respectable scientific journal *Physical Review* an article that suggested how to build a time-machine. It requires a cylinder 100 kilometres long and 10 kilometres wide – easy enough so far – but it has to be made of a material denser than a neutron star, which means very, very dense indeed, and, once that has been achieved, the cylinder needs to be rotated at a phenomenal speed of about two rotations per millisecond. That is not all. Once the cylinder is in rotation, the next thing is to steer a spaceship right up to it without either crashing or slipping into the holes at each end. Too complicated! It certainly puts time-travel out of the window for the time being at least. But modern science is alive to the idea that it might be possible in the future, if the technological advances of the last two centuries continue at the same rate of exponential development.

One of the greatest philosophical arguments against time-travel asks why, if people were able to go back in time, we have not seen or heard from them yet, invading our own time from the future. One bleak view suggests that, although human beings are theoretically able to travel backwards in time, the world will come to an end before we have technologically perfected our machine. Alternatively, if such a machine does exist in the future and is built, say, 200,000,000 years from now, is there any reason to suppose that time-travellers would wish to visit this particular narrow band of time in which we live? The machines would always be expensive and difficult to build; it is most unlikely that they would ever be a household convenience like the car. In which case there would be a limit to the number of people authorised to use the machine and the number of time zones they are therefore able to visit. To conclude this topic (which still seems babyish, in spite of the scientific backing which Tipler and his crowd have lent to it), it has also been suggested that while people can, theoretically, return to past time zones, they cannot be seen once they get there. Now we really are entering the fantasy world of the science-fiction writer. And this time it is not even new, as anyone who has read the opening paragraph of H. G. Wells' *War of the Worlds* (1898) will remember:

No one would have believed, in the last years of the nineteenth century, that human affairs were being watched keenly and closely by intelligence greater than man's and yet as mortal as his own; that as men busied themselves with their affairs they were scrutinised and studied, perhaps as narrowly as a man with a microscope might scrutinise the transient creatures that swarm and multiply in a drop of water.

CHAPTER XVI

FINIS
End

If we find the beginning of time difficult to comprehend, then the end of time is even worse. For many people it is repugnant to reason that time should vanish altogether. With it would go all trace of their lives, of anything that the whole human race has striven to achieve. There is a general acceptance that our world itself cannot go on for ever. But the end, we all assure ourselves, will never come in our lifetimes. Maybe human beings will be so technologically advanced by that stage that they will be able to desert the sinking ship – hop off the earth and escape on to another bigger, less crowded, more friendly planet. At the moment such a possibility lies only within the realm of science fiction. But what can we possibly say about the end of time that is not pure science fiction? Do we know anything about what will happen in the future, or can we only guess, cross our sweaty fingers, and hope for a desirable outcome?

When we look at time, and the way in which human civilisations have measured it, from the thuggish *Homo erectus* of yesteryear with his scruffily etched bones, to prototypical modern man thinking in precisely measured seconds, we see, in every direction, wheels and cycles, revolving cogs in a mechanical clock, the spheres of the earth and the moon, all the planets revolving around the sun, the ellipses and spheres endlessly returning in long unchanging circles, the South American Indian drawing a circumference with his arms to indicate the time of day, the hands of a clock ticktocking round and round. Why is everything round? Why does time have to be so cyclical?

Is it just us? Have we made time look round to soothe our silly anxieties? Perhaps it is our in-built fear of death, of linear time, driving forwards in one direction towards eternity, or an uncertain nemesis, that forces us, for our own sanity, to believe in cyclical time. But Reason tells us off. Time is irresistible, and it is irreversible, which is why man, since the dawn of civilisation, has attempted to rationalise it with theories of cyclical worlds and the eternal return. As Marcus Aurelius once wishfully boasted: 'All things from eternity are of like forms and come round in a circle.'

Even physicists, who are supposed to fly the flag of scientific impartiality, have, many of them, fallen for the comfort of cyclicity. One astrophysical theory holds that not only time, but space and the universe itself are indeed cyclical phenomena. As the universe continues to expand and the galaxies fly further and further apart from one another, the forces of gravity will by degrees retake the reins. Stars that have spent their nuclear energy will collapse inwardly, more black holes will be created, and into these will fall the dust and debris of particle matter, photons of light, quarks and leptons all vacuumed inwards by an irresistible force, a spinning vortex of drastically dense time and space, consuming itself and all around it into the frantic shapelessness of a zero-point singularity. From here the rest is timeless history. Bang! – and another universe is on the way.

Welcome to the bickering world of theoretical physics, where one professor's theory is claptrap to all the others and 1,000 rival theories abound.

It would be neat, unbelievably neat, if time and the universe were indeed cyclical, like almost everything else in our time consciousness, but at the moment even our most advanced scientists are a long way from discovering what lies at the bottom of our universal mechanism. Is there anything outside that is making it work? What is it that causes gravity to pull?

How will the world end? How can we ensure that life continues in the event of all our space-time disappearing up the back end of a black hole? All these questions need to be answered for the sake of life.

When we do discover the answers to some of them, it may be too late for us to take any action to save ourselves. But if we hold with fervour to any of the main religions we should be able to clench our teeth and smile like Marshall Herff Applewhite, the cult leader who died with his followers in the expectation of a coming nirvana. If we have been good that is; only then can we look forward to an apocalypse of *pâté de foie gras* munched to the sound of trumpets. If and when this ever happens, theoretical physicists will all be out of a job. Until then, however, we had better listen to what they have to say.

One of the tidiest theories of universal time (so tidy that it is generally dismissed) proposes a model of the universe which is spherical. 'Ho! ho! ho!' we all laugh. 'A spherical universe, that's rich!' Perhaps we should try to suppress our mirth for an instant. We know so little about the universe and how it works that almost anything in theoretical physics might be possible. Let us also not forget that Copernicus was laughed out of court in 1543 when it was discovered that his book *On the Revolutions of the Heavenly Spheres* had proposed that the earth orbited the sun. Nor should we forget that his ancestors – how many generations earlier? Not more than 50 – believed the world to be a flat disc off which one could easily fall. The advances in our knowledge since Copernicus first put pen to paper have been exponential, but this is not to say that what we now know, or at least hold to be true, is any greater than an atom in the face of all the knowledge that we have yet to acquire.

The spherical-universe theory holds that the big bang happened at the pole of a sphere. The explosion sent stars and galaxies shooting apart and covering the outer surface of that sphere in all directions. Having passed the halfway mark (the equator of the universe, as it were), all the flying matter will be pulled back together by the forces of gravity and will collide at the opposite pole. The intensity of such a huge gravitational collision will, we are told, result in a singularity that will duly explode into another universe of glowing, nuclear shrapnel. This, in turn, might or might not support life. In any event the whole lot of it will shoot back to the opposite universal pole from whence it originally came and the process can theoretically repeat *ad nauseam*.

Einstein's teacher, Hermann Minkowski, had, as we have seen, drawn space and time together as one unity called space-time. That was in 1908. Within months of his Cologne lecture every scientist worth his salt was working on space-time. Most important, it provided the mathematical backing for Einstein's general theory of relativity. Rumours emerged from Cologne, Göttingen and all the most eminent centres of astrophysical learning that space-time was curved and that a straight-line journey was now, officially, impossible. Science was leaving the amateur behind. According to one special theory, the furthest 'straight-line' distance from yourself in any direction would bring you back to yourself again; the universe was a geometrically four-dimensional sphere. In this and other models, physics was turning metaphysical. Einstein's relativity made so much possible that had never been possible before. It opened the floodgates for professionals and amateurs to indulge their wildest fantasies in the construction of bizarre theories, each attempting to explain the space-time geometry of the universe. In all this razzle, the truth somehow became less important than the question of egos. Fundamental questions like 'Who is going to be the next Einstein?' took precedence over such frivolous issues as 'Would time ever come to an end?' and 'Why is the universe hospitable to life?'

In terms of received opinion the professionals have had little difficulty in stamping out the models of their amateur rivals. In truth there are amateurs who are just as equipped as they. Let us not forget that Einstein was a patenting clerk when he came up with relativity. The Internet is now the festering centre for amateur speculation about the universe. The search is on for the holy grail of physics – a Grand Theory of Everything (TOEs as they are known to the Internet boffins, or, less pleasingly, as GUTs, which stands for Grand Unified Theories) that can be represented in a simple mathematical equation. Might such a TOE or GUT one day surface from the depth of cyberspace? Maybe it is there already, swimming in the ether. 'Full many a flower is born to blush unseen and waste its sweetness on the desert air.'

There are scientists who hold that our universal time, once created, is eternal. Parts of it might get sucked up into the whirligig of a black hole, but threads will continue for ever. These scientists are supporters of the theory of 'entropy', which is dictated to us by the natural consequences of Rudolf Julius Clausius's second law of thermodynamics. An explanation is due. But first let us hear what Arthur Eddington (1882–1944), father of modern astrophysics, had to say on the matter in his book *The Nature of the Physical World*:

> The second law of thermodynamics holds, I think, the supreme position among the laws of Nature. If someone points out to you that your pet theory of the universe is in disagreement with Maxwell's equations – then so much the worse for Maxwell's equations. If it is found to be contradicted by observation – well, these experimentalists do bungle things sometimes. But if your theory is found to be against the second law of thermodynamics I can give you no hope; there is nothing for it but to collapse in deepest humiliation.

So what is the second law of thermodynamics? Who was Clausius? And what has this all got to do with the end of time?

Rudolf Clausius (1822–88) was a Prussian, a mathematical physicist from Köslin. Apart from the inclusion of a long list of professorships starting at Zurich in 1855 and ending at Bonn in 1869, Clausius's biography is not interesting enough to precis. Let us then content ourselves with a flying description drawn from a photograph of Clausius taken in 1880, when he was at the height of his fame – mean, zealous, self-regarding, a beard without moustache, a pinched lower lip, balding on top and a fixed gaze in between.

Now, the second law of thermodynamics states what might appear to any non-scientist to be so obvious that it hardly needs laying down as a scientific law, and certainly does not deserve to be sanctified by the likes of Arthur Eddington. The law simply states that things wear out. There is, it would appear, a one-way time direction in which things dilapidate. It is called Time's Arrow. A vase might break if it falls to the ground or if crashed into by a clumsy teenager, but it will never reassemble itself of its own volition because time moves in one direction only and physical objects tend to observe the rule of Time's Arrow. Time wears things down and breaks them up into smaller parts. The second law of thermodynamics is often expressed in terms of heat. Heat flows from a hot object to a colder object so that both objects, if left to their own devices, will eventually reach a state of thermal equilibrium, but the process will never happen the other way around. (According to the French physicist Henri Poincaré, cold could flow back into heat if we gave it infinite time to do so – we shall ignore Poincaré's point for the time being.) If, like John Harrison, you set two adjoining rooms of your house at different temperatures (one piping hot and the other freezing cold), when you open the adjoining door everyone knows what will happen. The heat will flow out of the hot room into the cold room and in due course both rooms will record the same temperature. But the heat, however little there was in the colder of the two rooms to start with, will never pass from the cold into the hotter room.

This one-way process, in which things are said to move from order to disorder and from heat to thermal equilibrium, is known to physicists as 'entropy'. The most fundamental laws of physics state that the entropy of a closed system always increases. That is the scientists' way of saying that things always wear out in the end. If we think of the universe as a closed system, then what this means is that all the matter in the universe, from the stars and planets to your computer and your bed, will eventually disintegrate. Stars will burn out and their heat will diffuse into space. Such heat will never return to the stars from whence it came. It will simply move to fill colder areas of the universe so that the whole universe will eventually reach thermal equilibrium at a temperature that would be far too cold to support any life forms. As to the stars themselves, the matter from which they are made will eventually fall apart, and its atoms will be spread evenly across the whole universe – they will never, according to this theory, reassemble back into the original or into any other star. Similarly, when a vase is dropped on the floor, its kinetic energy is transferred into heat, some of which will dissipate by way of infra-red radiation into space and will never be heard from again. Everything has energy, either by way of mass (Einstein showed that mass and energy were interchangeable in the equation $E=mc^2$) or kinetic energy, which is expressed as motion. If Clausius's second law of thermodynamics is right and really is as irrefutable as Arthur Eddington says it is, then there is only one future for this universe of ours. Everything must fall to pieces and spread itself evenly across space. The universe itself will gradually settle as an unfathomably vast, four-dimensional surface in which all particles of matter will be equally distributed at an unbearably low uniform temperature. This theory is called the 'heat death' theory of the universe. In such a model, time would come to an end when the universe reaches a final static state of equilibrium.

In February 1999 a 10-year plan to map a 500-million-square-mile section of the universe was finally unveiled to the public. To the layman it looked like a surreal collection of floating potatoes, but the Anglo-German team which created it vehemently protested that their map proved, beyond reasonable doubt, that our universe will carry on expanding for ever. But hadn't they forgotten something? What about the heat death theory?

They probably chose to ignore it, as many scientists before them have done, because there is a problem which the heat death theory fails to acknowledge. It is the problem of life. Life regenerates itself and makes things like vases. On every inch of the earth's surface, where there is light, there is some form of life. It is as though the planet itself were alive. In this way the existence of life acts as the opposite force to the physical law of entropy. Some have argued that if there were enough life in the universe as a whole, Clausius's hitherto irrefutable second law of thermodynamics might somehow be reversed. As with so many important issues of theoretical physics, scientists are still quarrelling over this one – but, if life was ever *proved* to decrease entropy of the universe, Sir Arthur Eddington and his friends might themselves be forced to 'collapse in deepest humiliation'.

According to the theoretical physicist Lee Smolin, the whole universe is organised along Darwinian lines. By this he implies that it is evolutionary and self-correcting, built not by one God but formed, like a city, by and from the intricate interrelation of its parts. For Smolin there can be no end to time:

> So there never was a God, no pilot who made the world by imposing order on chaos and who remains outside, watching and proscribing ... The eternal return, the eternal heat death, are no longer threats, they will never come, nor will heaven. The world will always be here, and it will always be different, more varied, more interesting, more alive, but still always the world in all its complexity and incompleteness. There is nothing behind it, no absolute or platonic world to transcend to. All there is of Nature is what is around us. All there is of Being is relations among real, sensible things.

The more religious you are, the more irritating you will find Lee Smolin's views. Most religions hold that the end of earthly time will come in some form of apocalypse, an event which will clear out the odium of evil in a fantastic courtroom drama (the Last Judgement) preceded by a pitched battle between Satan (chaos) and God (cosmos), which will be followed by the destruction of all that is worldly and the restoration of the original timeless paradise of God's kingdom. The meek shall be elevated to an astral plane, as guests of the Almighty, supping at his table, drinking the elixir of eternal life, making light, inoffensive jokes and wandering, after lunch, through the restored Garden of Eden to take in the exquisite scenery and fine outdoor smells. Jesus promises these things and more besides.

If, as the old saying goes, the way to a man's heart is through his stomach, Jesus certainly offered the right bait to his followers. He was keen on food and wine himself. Let us not forget how he turned the water into wine at the marriage feast at Cana, how he fed the 5,000 with fish and bread and how he cursed the fig tree that failed to produce fruit out of season.

It is a wonder that Jesus was so thin, for food was never far from his thoughts and, even after his death, he continues to transubstantiate himself into bread and red wine for the benefit of Roman Catholics every Sunday. In the New Testament, Jesus promises his virtuous followers a whole range of post-apocalyptic eating opportunities. He is also full of angry vengeance at those who might not give him food: 'Depart from me, ye cursed, into everlasting fire, prepared for the devil and his angels: For I was an hungred, and ye gave me no meat: I was thirsty and ye gave me no drink.' The same metaphor was present in his address from the Mount of Olives: 'Blessed are they which do hunger and thirst after righteousness: for they shall be filled.' His followers are promised that God's kingdom will have plenty for them to eat and drink. 'I tell you this,' he says at the Last Supper, at which he 'dippeth his hand into the dish' and consumes fish and unleavened bread and red wine, 'never again shall I drink from the fruit of the vine until that day when I drink it new in the kingdom of God.'

But Jesus was not by any means striking a new line here. The first promise of food in the kingdom of heaven is suggested way back in the Old Testament Book of Exodus. The kingdom of God is a 'good and broad land, a land flowing with milk and honey.' Later, in the Book of Isaiah, God's eternal supper sounds even more delicious than that:

On this mountain shall the Lord of Hosts make unto all people a feast of fat things, a feast of wines on the lees, of fat things full of marrow, of wines on the lees well refined.

If you break any one of Moses' Ten Commandments you are branded a sinner, and if you subsequently die unrepentant you will be cast into eternal darkness. We have seen how God whispered to Moses to ask the 'children of Israel' to stone a helpless man in the wilderness for picking up twigs on a Sabbath. Christianity is a tough religion, let us not forget it. The Old Testament is just as frightening and barbaric as anything in Islamic fundamentalism. Yet the rewards for the good Christian are indulgent in the extreme. Bishop Papias of Phrygia in the first century AD, who went around collecting stories about Christ from people whose parents might have known him, wrote at length about the promise of delicious grapes and bread which he himself was greatly looking forward to after the coming millennium:

> The days will come in which the vines shall appear, having each ten thousand shoots, and on every shoot, ten thousand twigs, and on each true twig ten thousand stems, and on every stem ten thousand bunches, and in every bunch ten thousand grapes, and every grape will give five-and-twenty gallons of wine. And when any one of the saints shall take hold of a bunch, shall cry out, 'I am a better bunch, take me, bless the Lord through me.' Likewise the Lord saith that a grain of wheat would bear ten thousand ears, and every ear would have ten thousand grains and every grain would give ten tons of the finest flour, clear and pure; and apples and seeds and grass would produce in equal proportions, and all animals feeding only on what they received from the earth, would become peaceable and friendly to each other, and completely subject to man. Now these things are credible to believers. And Judas, being a disbelieving traitor, asked, 'How shall such growth be brought about by the Lord?' But the Lord answered, 'They shall see who shall come to those times.'

The Book of Revelation, traditionally believed to be the work of St John the Evangelist, Jesus's close friend and the first disciple to be chosen, points to other benefits. God assures his followers that he 'will give unto him that is a thirst of the fountain of the water of life freely . . . and there shall be no more death, neither sorrow nor crying, neither shall there be any more pain: for the former things are passed away.'

In the Koran, the most tempting paradise of all is offered to the devout disciples of Allah:

> But the true servants of Allah shall be well provided for, feasting on fruit, and honoured in the gardens of delight. Reclining face to face upon soft cushions, they shall be served with a goblet filled at a gushing fountain, white and delicious to those who drink it. It will neither dull their senses, nor befuddle them. They shall sit with bashful, dark-eyed virgins, as chaste as the sheltered eggs of ostriches.

No wonder these religious folk do not wish to hear Lee Smolin's rant about our 'interesting', 'alive', 'eternal' world. For the good apocalyptic Jew, for the Christian or the worthy Muslim, our flawed, chaotic world must come to an end so that the eternal beano of grapes and grain, soft cushions and bashful virgins can begin in the eternal kingdom of heaven. To the cynic there is something startlingly patronising about all this. Even the aforementioned Papias's little lecture cannot fail to echo the way parents bring up their children: 'If you don't tidy your room, there won't be any pudding for supper.' But such nanny-state connections are missed by the believer.

To the Buddhist the notion that a human being has only one life on earth is judged to be an illusion borne of lost memory. Buddhists have all had many lives and will have many more if they are not virtuous enough to escape the eternal wheel of life, death and rebirth. If they are surly, bad-mannered or sinful, they will be reborn as worms or lower life forms; if they behave themselves they will ascend in the chain. This is supposed to obliterate the fear of death, but it does not. Rebirth might be a belief, but it is still the unknown. The Buddhist's main objective is to escape the 'sorrowful wheel' (the dukkha) and move on to another plane, a transcendental state known as Nirvana, where once again earthly time is ended and a new, unbounded time begins. In 1999 the England team football coach, Glenn Hoddle, expressed, as his own belief, the ancient Pythagorean view that handicapped people are atoning for the sins of their previous existence. The Prime Minister, Tony Blair, pronounced in an interview on daytime television that Hoddle should be sacked for talking in this way, and within days Hoddle was out of a job.

Through Tony Blair's Christian eyes, the ancient Egyptian system of immortality must look as offensive and undemocratic as Hoddle's transmigration of souls. But Mr Blair is not in a position to relieve the ancient Pharaohs of their posts. In Egypt labourers (who feared death just as much as their kings did) were told that they could continue after death indefinitely as labourers, working and struggling in the same manner as on earth – only if they were good, of course. The Pharaoh, however, had a better deal; he could drift off with Ra in the sun-god's plush boat, taking as many of his friends and relations as he chose to live with him on the sun. From there he was believed to continue protecting the land and the people of Egypt for evermore. But Egyptian gods, and their relations, had to be constantly fed. It was the same in Mesopotamia. Enormous quantities of lavish food were supplied to the temples day after day. This included bulls, boars, sheep and fowl, flagons of wine, beer, junket and bread by the tonne.

Such a curious connection between timelessness, the afterlife and food was therefore a live tradition thousands of years before Christ. In 3300 BC the Egyptian *Book of the Dead* described the 'other world' as a place of cakes, cool water, ale, fish and delicious milk to drink. Whether this was referring to the food that the people of Egypt were leaving at the doors of the temples, or to some other food which the dead could enjoy in addition to the sacrificial piles, is unclear.

The Buddhist 'happy land' is not so concerned with food as the others, but there are attractive smells there. There are beautiful birds as well, and flowers with precious jewels stuck to them. There is also a river, a mighty flux 50 miles (80 kilometres) broad and 12 miles (19 kilometres) deep, which makes beautiful music as it flows. The description is ironically close to the earthly paradise from which the great Buddha fled in his thirtieth year to lead the humble life of an ascetic.

St Augustine of Hippo taught his doctrine of 'two cities' in which the battle of good and evil had already been fought and won (by God, of course) but not so conclusively that Satan could not stay around in his own city for the time being. In the present age the City of the World and the City of God were forced to coexist. Eventually, however, Satan would lose his patronage of his world and God would be triumphant.

But this is all cold comfort to the millennial panicker, scared stiff that all of time is about to come to a sudden and irresistible full stop, snuffing out his wisdom and his life. If it is not God who is going to bring it all to an end, might it be the devil? Will we all be wiped out by AIDS or another deadly virus? Or will we just crowd ourselves off this tiny, stinking planet? In November 1992, popular newspapers ran with headline banners proclaiming 'THE END OF THE WORLD!' They were referring to recent research by cosmic scientists into a comet called Swift-Tuttle which, they were earnestly promising, would collide with the earth in August 2016, bashing our tiny planet out of its orbit with the sun and casting us all into eternal freezing cold darkness. What an unpleasant end, and how undignified for the whole of earthly life to be extinguished by something with a name as bizarre as 'Swift-Tuttle' – let us at least change it to 'Vulcan Avenger' or the 'Beast from Hell'. But haven't we heard it all before?

One form of street attire among the eccentric minority communities of our large cities is the sandwich board: two heavy wooden placards strung over the shoulders of an importunate proselytiser, bearing messages or trade advertisements on either side. 'Jesus says this' or 'Jesus says that' – most commonly they are quasi-religious warnings that time is running out for the sinners of the world. 'The End is Nigh' is a regular message which seldom fails to draw attention. But in ancient Rome such a parade would have been prohibited. Any speculation about how the gods may or may not choose to deal with you once you are dead was considered out of bounds, as Horace warns: *Tu ne quaerieris, scire nefas, quem mihi, quem tibi finem diderent.* ('Do not try to find out, for we are forbidden to know what end the gods have in store for me or you.')

When the English painter William Hogarth was asked what would be the subject of his next print at a party in 1764, he replied: 'The End of All Things.'

'If that should be the case,' replied his companion, 'your business will be finished, for there will be an end of the painter.'

'There will so,' said Hogarth, 'and therefore the sooner my work is done, the better.'

Is it then a strange irony that Hogarth's last picture should have been a mysterious representation of the end of time? Like Mozart in his mid-thirties, who died while composing a requiem Mass which he had come to believe was a signal to his own death, Hogarth, too, worked furiously at his 'tail-piece', convinced that the picture would mark a 'terminus to his fame'. Indeed, it did, for he died as soon as it was finished. *The End of All Things* is an extraordinary picture, filled with the symbols of time and decay as they were recognisable to educated classes of mid-eighteenth-century England. On one side is a ruined tower with a broken clock on it; nearby a gravestone carved with the skull and crossbones. The central figure is of Old Father Time himself, ancient, collapsed and winged, lying in a heap against a brick column. In one hand he holds a broken pipe, in the other his last will and testament, leaving every atom of this world to blank chaos. His will is witnessed by the three sister Fates – Clotho, Lachesis and Atropos. In the sky the moon is darkened by the death of Sol, who lies dead on a broken wheeled chariot, as his horses, dead as well, tumble from the sky. Other symbols in the picture include Time's famous hourglass showing that time has run out, a bankruptcy order taken out against nature, a cracked bell, a broken pub sign for the World's End depicting the earth in flames, a withered tree, a broken crown and, in the far distant background, a hanged body swinging from the gallows.

And so, as the popular nineteenth-century editions of Hogarth's *Works* concluded:

> However jumbled may be the objects in this plate, with the design of exposing the absurdities of some ancient paintings, they serve to put us in mind that life is little better than a jumble of incidents; that the end of all things approaches; and that a day will, sooner or later, come when Time itself shall be no more.

In music, the finest tribute to the end of time is Olivier Messiaen's *Quator pour la Fin du Temps* (Quartet of the End of Time). It was composed in a prisoner-of-war camp called Stalag VIII A, near Görlitz in Saxony. The first performance of this extraordinary, intimate piece took place in front of 5,000 French, Belgian and Polish prisoners on 15 January 1941. Messiaen took his initial inspiration from the Book of Revelation:

> And I saw another mighty angel come down from heaven, clothed with a cloud and a rainbow was upon his head, and his face was, as it were, the sun and his feet as pillars of fire. And he had in his right hand a little book open: and he set his right foot upon the sea, and his left foot upon the earth . . . And he sware by him that liveth for ever and ever, who created Heaven and the things that therein are, and the sea and the things which are therein, that there should be time no longer.

The composer was fundamentally interested in exploring the ways in which music could be used to break down our concepts of time, using long held notes and sustained sounds or irregular rhythms that walk free from the normal pulse of Western music. In this he was influenced by the sounds of India and the Orient.

The *Quator pour la Fin du Temps* is an exquisite work that gives a feeling of the end of time which words cannot possibly hope to convey. It is scored for piano, violin, cello (with one string missing – for that is how it was in Stalag VIII A) and a single clarinet. Never miss an opportunity to hear this piece played live.

If the *Quator pour la Fin du Temps* may have tragic as well as ethereal overtones (people stagger out of performances of this work as though in a coma of sublime shellshock) it is because Messiaen, a deeply religious man, approached the Apocalypse with a sense of wonderment – a feeling which has nothing to do with the breathless millennial hysteria that we see elsewhere.

Mass ritual suicides by cult followers in the late twentieth century have provided us with evidence of the most sinister and selfish side of apocalyptic belief. David Koresh's cult was assaulted by a trigger-happy FBI in Waco, Texas. In California, shortly afterwards, 39 members of the Heaven's Gate cult were found dead, neatly laid out on beds with bags over their faces. They were all followers of Marshall Herff Applewhite, a retired music teacher who had taught them that they were taking the first step in a millennial flight to the heavenly 'level above human' and that their transport was to be a spaceship trailing in the wake of the Hale-Bopp comet which appeared in the night skies at that time. Five of the men, including Applewhite, had castrated themselves as part of a cleansing ritual that involved total abstinence from sex and drugs.

On the day before their deaths, they watched *Star Trek* on television, filled themselves with hamburgers and pizzas, and each made a short video explaining his or her happiness at departing this doom-laden world for something better. During this period, fuelled by the belief that Applewhite was dying of cancer, the members put on uniforms of black trousers, shirts and new Nike training shoes (ironically emblazoned with the white comet-like trademark) and ate a concoction of apple sauce and pudding mixed with phenobarbital and washed down with vodka. Putting plastic bags over their heads, members settled down on 'comforters' with five-dollar bills in their pockets and waited for death. For Heaven's Gate, time was running out. They were leaving the rest of us behind to die our agonising deaths, crushed by fiery boulders and squeezed between two enclosing plates of time. The Heaven's Gate followers, on the other hand, were smug in the belief that they had caught the first, the last and the only bus out of town, Hale-Bopp.

The Hale-Bopp bus took them nowhere. It is easy to suppose that the Heaven's Gate members were just mixed-up loonies or, according to one book on the matter, show-offs who died for the sake of entertaining the world with their story. But the Heaven's Gate fears were ancient fears, and so were their aspirations. Modern cynicism has fortunately taught most of us to dispel such thoughts as mythological hocus-pocus. According to the fiercest tenets of Christian faith, those 39 Heaven's Gate members will not have got off so lightly. For a start, suicide is a sin and there is always the Last Judgement to come, when God will weigh up both the living and the dead, before deciding on whether or not to 'call them to his supper'. In the early centuries after the life of Christ there were undoubtedly more gullible people around than there are nowadays, and the scriptures taught them to aim for the grapes or to suffer the consequences.

Can this explain Marshall Applewhite's castrated rush to catch the passing Hale-Bopp? The alternative to God's kingdom is far more unpleasant than 'no pudding for supper'. Revelation is expressly clear about this: 'He that overcometh shall inherit all things', but he who overcometh not . . .

The fearful, and unbelieving, and the abominable, and murderers and whoremongers, and sorcerers, and idolaters, and all liars, shall have their part in the lake which burneth with fire and brimstone: which is the second death.

This, for centuries, has survived as the acceptable face of protectionism, whose uglier side was manifested in the Kray twins, 1960s hooligans who extorted money from shopkeepers in the

East End of London for 'protecting' a shopkeeper's property (i.e. not burning it down). On the basis of the doctrine of Revelation, the Kray twins, unless they have fully repented, will be heading for the lake of fire and brimstone. But you do not have to be as wicked as the Kray twins to attract God's opprobrium. God makes all the rules, and he, at the Last Judgement, will lock 99.9 per cent of the human population (liars and non-believers) into an eternity of scorching, sulphurous lava before repairing to the dining room with a paltry following of sycophantic believers. We hope they gag on their grapes!

There are problems with hell which the architects of religion may have failed to spot. The proportion of damned to saved is very uneven. Of course, the 'chosen ones' are the elect, and there is no point in having an elect if the majority is chosen to be in it, but, even so, casting more or less everyone into hell and allowing just a few into heaven cannot be a sensible procedure for a God who is supposed to be just, wise and good. As Leibniz pointed out, if vastly more people end up in hell than in heaven, then the universe will have more evil than good in it, which cannot have been God's original intention:

> It seems strange that even in the great future of eternity, evil must win over good, under the supreme authority of Him who is the sovereign good; for there will be many called and few chosen or saved.

He is referring to the Book of Matthew: 'For many are called, but few are chosen,' or, as he makes clear earlier on:

> Enter ye in at the strait gate: for wide is the gate, and broad is the way, that leadeth to destruction, and many there be which go in thereat: Because strait is the gate, and narrow is the way which leadeth unto life and few there be that find it.

So almost everybody sweeps through the wide gate and finds himself in hell, even unbaptised infants. So many souls that, according to one theory, hell must be in the sun, 'for there could not possibly be room in the bowels of the earth for so many'. One German theologian, Drexelius, calculated that the damned were packed in at 100,000,000,000 to the square mile. Tobias Swinden, in his *Enquiry into the Nature and Place of Hell* (1714), comments: 'It is a poor, mean and narrow Conception both of the Numbers of the Damned, and the Dimensions of Hell, which Drexelius hath laid down.' As far as the doctrine of unbaptised babies is concerned, Tobias Swinden has no comfort to offer their aggrieved and mourning parents: 'the subject of it, although melancholick and sad, is yet beyond a possibility of Contradiction.'

At times when the mortality rate among the newborn was exceedingly high, it could be said that 50 per cent of the population of hell was made up of babies. Catholics, who were on the whole uncomfortable about this, consigned the babies not to hell but to limbo, where they were not tortured but simply deprived the pleasures of the beatific vision.

If the biblical hell offers an unpalatable view of the end of time, the Muslim Koran is no more comforting:

For the unbelievers we have prepared fetters and chains and a blazing fire. But the righteous shall drink of a cup tempered at the Camphor Fountain, a gushing spring at which the servants of Allah will refresh themselves.

Islamic hell is called *jahannam*. To get to heaven you must cross a razor-thin bridge. If you fail to get across, because you have sinned too much in your earthly life, you will fall into the concentric circles of *jahannam* where you will be punished according to the gravity of your offences and released only when and if God wills it. One odd twist of the Islamic hell is that you are allowed to take with you all your lusty thoughts, but there will be no bodies there for you to fulfil your desires.

The word *jahannam* is derived from the Jewish Gehenna, which in turn refers to Ge Hinnom (the Valley of the Hinnom). This is a real place to the south-west of Jerusalem where for more than seven centuries at Topheth (from the time of Moses in the fourteenth to thirteenth centuries BC to the Babylonian exile in the sixth century BC) the Israelites burned their children on blazing bonfires as sacrifices to the Ammonite god Moloch. Moloch, after whom the hideous Australian ant-eating sand lizard (*Moloch horridus*) is named, derives his own name from mixing the Hebrew *melech* ('king') with the vowels of boshet ('shame'). Moses, in one of his more humane moods, forbade the Israelites to throw their children on to the fire for Moloch.

And thou shalt not let any of thy seed pass through the fire to Moloch. Whosoever he be that giveth any of his seed unto Moloch; he shall surely be put to death: the people of the land shall stone him with stones. And if the people of the land do anyways hide their eyes from the man, when he giveth of his seed unto Moloch and kill him not, then I will set my face against that man, and against his family, and will cut him off, and all that go a whoring after him, to commit whoredom with Moloch.

Old habits unfortunately die hard. For all Moses' injunctions, Israelite children continued to be thrown on to the fires at Topheth for another 700 years, until the reign of Josiah (*c* 640–609 BC) when the reforming King 'defiled Topheth which is in the valley of the children of Hinnom, that no man might make his son or his daughter to pass through the fire to Moloch'. After Josiah's defilement Ge Hinnom was turned into a rubbish dump with a view to making it so stinky and infectious that the Israelites wishing to sacrifice their children there would choose to stay away rather than brave the awful smell. The ruse worked. Ge Hinnom, the once-popular infant sacrifice point, became Gehenna, a deserted area which the Jewish finally associated with hell. It is from this idea of Gehenna and the imagery of burning infants that the concept of 'hellfire' is supposed to have derived.

There is something very basic about this all-or-nothing attitude to religion. Vicious torture or comfort and food but nothing much in between. In a curious piece of sophistry the seventeenth-century mathematician, Blaise Pascal, argued a rational argument for believing in God based on a wager. 'If God exists and you spend your life believing and praying to him, you will end up, for the rest of eternity, in the land of milk and honey. If He does not exist – well, you haven't lost much, have you? If, on the other hand, God *does* exist and you have not been believing and praying to Him, then you are in for an eternity of fire and brimstone. Therefore,' according to Pascal's logic, 'unless the probability of God's existence is infinitesimal, you would be better off believing in Him.'

There are a number of things wrong with this argument. On the one hand it supposes that God (who is supposed to be omniscient) would be happy in the knowledge that your belief in him was based on gambling odds, and on the other hand it assumes that God really is so vain and oppressive that he will leave non-believers in an eternally bubbling vat of oil while handing out beakers of milk and pots of honey to his sycophantic believers. God might exist and want to be believed in for honourable reasons that are not to do with gambling odds. Alternatively, he might not give a fig whether people believe in him or not. If God has a sense of humour (another possibility that Pascal had not considered), he might choose to drop the believers into the vats of boiling oil and go off for supper with the whoremongers – who knows? God-fearing Christians believe they do know, and it is because they *know* that they are so eager to earn a place for themselves in the kingdom of heaven after the Last Judgement.

When asked where he wanted to end up after his death, Oscar Wilde replied: 'Both. Heaven for the scenery, hell for the company.'

Whether Wilde is now looking at the scenery of God's kingdom must remain uncertain. The puritan Christians would gnash their teeth at the suggestion of any milk and honey going Wilde's way, for the holy are quick to damn the sinners. Their argy-bargy to get to heaven is the result of a panic, based on certain scriptural passages that confirm that the University of God has not got many places to offer. To this the punishment-loving puritan may add his own peculiar hatred of mankind. 'Hell was invented by men with resentment,' said Freud, and evidence has shown that the people most obsessed with eternal damnation are often those groups which are living in a state of extreme misery and oppressed by tyrants whom they cannot rationally hope to throw off. Their anger and frustrated vengeance find an outlet in the hope for a Messiah who will come down to earth and clobber their enemies, casting them all into eternal darkness. Extremism was at its height in the Middle Ages. It had a second wind in the seventeenth century. In England Dr Richard Roach taught the Gospel of God's Love and found himself up against the Camisards, whose belief was not in divine love so much as divine justice – love did not come into it. At one meeting, a Camisard prophetess, Mary Keimer, stood up, blue with rage. For this malevolent woman, God was not a God of love but a God of justice, and anyone who did not agree with her should join the millions whose bodies are set to rot in the mire-pit of eternal hell.

'I will confound thee,' screamed Mary Keimer at the hopeless, sanctimonious Roach:

Yea, and bring thy lofty thoughts down; I will lay thee even with the dust. Who art thou, O man that Exaltest thyself? Who art thou? I *am*, I *am*! How durst thou presume to speak unto me? I can this moment strike thee dead . . . I *am*, I *am*, I *am*, I *am*, I *am*, . . . and ye shall know that I have spoken & who now does speak.

This was not an isolated incident of Mary Keimer's wrath. She was much to be feared. According to her own prophesy she would go to France and speak to the king, 'who should upon his Disobedience be immediately struck dead by her mouth'. Her brother, Samuel Keimer, made her out to be stark staring mad:

I have seen my sister, who is a lusty young woman, fling

another Prophetess on the Floor, and under Agitations tread upon her Breast, Belly, Legs etc. walking several Times backwards and forwards over her, and stamping upon her with violence. This was adjudg'd to be a sign of the fall of the *Whore of Babylon*.

What an absurd performance; but this is the nature of fanaticism. Aldous Huxley was not just aiming at Mary Keimer when he wrote: 'At least two-thirds of our miseries spring from human stupidity, human malice and those great motivators and justifiers of malice and stupidity, idealism, dogmatism and proselytising zeal on behalf of religious or political idols.'

If fear of the afterlife, the stampede to squeeze through the narrow gates into heaven and avoid the broad entrances to hell, was a driving factor behind so many of the retributive religions (Christianity, Judaism and Islam), it played a surprisingly weak role in the pantheistic religions of ancient Rome. For the Romans the afterlife was a much more matter-of-fact affair. They minded more about the life that had been allotted to them on earth, than the life they may or may not enjoy after death, and to that end they made sacrifices to their gods, praised them, celebrated their festival days, and in doing these things hoped that the gods would, in turn, reward them with good health, prosperity and general happiness during their life on earth. However things turned out, for better or for worse, the superstitious ancient Roman always held the gods to be responsible. But when death came, that was the end; to suggest otherwise was a blasphemy. Only the gods were immortal. To intelligent Romans time was snuffed out at the moment of death, as Seneca explains in the *Troades*:

> After death nothing is, and nothing death:
> The utmost limit of a gasp of breath.
> Let the ambitious zealot lay aside
> His hopes of heaven, whose faith is but his pride;
> Let slavish souls lay by their fear,
> Nor be concerned which way or where

After this life they shall be hurled.
Dead, we become the lumber of the world,
And to that mass of matter shall be swept
Where things destroyed with things unborn are kept.
  Devouring time swallows us whole;
Impartial death confounds body and soul.
  For Hell and the foul fiend that rules
  God's everlasting fiery jails
  (Devised by rogues, dreaded by fools),
With his grim, grisly dog that keeps the door,
  Are senseless stories, idle tales,
    Dreams, whimseys, and no more.
          tr. John Wilmot, Earl of Rochester, 1680

To the practical mind the end of universal time is something which will never be experienced. The only end of time that can be experienced is the moment of our death. It is believed that the human being is the only member of the animal kingdom able to conceptualise past, present and future and consequently the only being which is aware of the processes of change, ageing, decay and ultimately death, which inevitably affects all living things.

For this reason human beings are uniquely obsessed with death, with mortality and with the end of time. William Hazlitt (1778–1830) had a comforting word:

Perhaps the best cure for the fear of death is to reflect that life has a beginning as well as an end. There was a time when we were not. This gives us no concern – why then should it trouble us that a time will come when we shall cease to be? I have no wish to have been alive a hundred years ago, or in the reign of Queen Anne: why should I regret and lay it so much to heart that I shall not be alive a hundred years hence, in the reign of I cannot tell whom? ... To die is only to be as we were before we were born; yet no one feels any remorse, or regret, or repugnance, in contemplating this last idea. It is rather a relief and disburthening of the mind:

it seems to have been holiday-time with us then: we were not called to appear upon the stage of life, to wear robes or tatters, to laugh or cry, be hooted or applauded; we had lain *perdus* all this while, snug, out of harm's way; and had slept out our thousands of centuries without wanting to be waked up; at peace and free from care, in a long nonage, in a sleep deeper and calmer than that of infancy, wrapped in the finest and softest dust. And the worst that we dread is, after a short, fretful, feverish being, after vain hopes, and idle fears, to sink to final repose again, and forget the troubled dream of life!

Hazlitt's idea is indeed a comforting one. If the state of not-being is impossible to imagine, it must also follow that it is impossible to fear. But logic is one thing, suppressing one's fear of timelessness, that great unknown, is quite another, as Tom Stoppard's neurotic monologue from *Rosencrantz and Guildenstern are Dead* makes abundantly clear:

Do you ever think of yourself as actually *dead*, lying in a box with a lid on it? . . . It's silly to be depressed by it. I mean one thinks of it like being *alive* in a box, one keeps forgetting to take into account the fact that one is *dead* . . . which should make a difference . . . shouldn't it? I mean, you'd never *know* you were in a box, would you? It would be just like being *asleep* in a box. Not that I'd like to sleep in a box, mind you, not without any air – you'd wake up dead, for a start, and then where would you be? Apart from inside a box . . . because you'd be helpless wouldn't you? Stuffed in a box like that, I mean you would be in there forever. Even taking into account the fact that you're dead really . . . I wouldn't think about it if I were you. You'd only get depressed. (*Pause.*) Eternity is a terrible thought. I mean, where's it going to end?

Are these paradises all just empty dreams? Is the mere thought of a finite time so terrible that we have to imagine ourselves in an eternal realm with bashful virgins and pots of honey? The basis for religious doctrine is a fascinating subject, for the roots of it go so far back in time, right to the dawn of the Sumerian civilisation and probably beyond that. It has become quite impossible to tell if the dogma was written of the people, by the people or for the people. Certainly it has been said that the one great benefit of religion is that it enables the controllers of it to organise the people around them more effectively. By invoking the word of God, kings, emperors, Jacobins, zealots and cult leaders of all shapes and sizes have managed to boss the gullible into doing almost anything they had in mind. Karl Marx and Friedrich Engels in their *Communist Manifesto* (1848) showed contempt for the religious utopia, but Bertrand Russell, that cunning old fox, was quick to point out some parallels between Marxist Communism and the Christian schema. He compared the 'materialist dialectic of the Marxist scheme to the biblical God, the proletariat to the elect, the Communist Party to the Church, the revolution to the Second Coming, and the Communist commonwealth to the millennium'.

For all the good that has come out of religion, there has been an equivalent amount of destruction, torture and odium. In terms of the answers to all our questions, scientists, philosophers and religious leaders are all prepared to offer a bewildering number of contradictory solutions. Until there is any conclusive proof, the individual is surely licensed to decide for himself. For many, the scientific black hole is no more credible than the religious utopia, but in the modern age of healthy scepticism it is always worth stopping to ask: who is telling me this and why?

'Sire, I have been mighty cunning today. I have written it down, on yon tablet of stone, that the meek shall inherit the earth.'

'What? We can't go giving the earth to the meek. How dare you, man!'

'Precisely, sire, but have no fear. The meek shall not inherit a brass farthing. I shall see to it myself. I have only written it on this tablet so that the people shall believe it, and you, sire, shall be able to boss them about with greater efficacy than ever before.'

'How so?'

'For it is written in tablets of stone, sire. Behold! I wrote them myself.'

'Excellent, Jehoshaphat, bravo! Have some more milk and honey.'

PICTURE CREDITS

Page 1: *top* Index/Bridgeman Art Library, *bottom left and bottom right* Mary Evans Picture Library

Page 2: *top and bottom* Sheridan/Ancient Art & Architecture Collection Ltd

Page 3: *top left* Sheridan/Ancient Art & Architecture Collection Ltd, *top right and bottom right* AKG Photo London, *bottom left* The Louvre, Paris

Page 4: Bridgeman Art Library/Museé Condé, Chantilly

Page 5: *top* Bridgeman Art Library/Christie's Images, *bottom* Bridgeman Art Library/Wallace Collection, London

Page 6: *top* Bridgeman Art Library/Phillips, *bottom* Mary Evans Picture Library

Page 7: *top* E.T. Archive, *bottom* AKG London

Page 8: *top left, top right and bottom* Hulton Getty

If you enjoyed this book here is a selection of other bestselling non-fiction titles from Headline

JAMES HERRIOT	Graham Lord	£6.99	☐
RICHARD BRANSON	Mick Brown	£8.99	☐
THE GIRL FROM LEAM LANE	Piers Dudgeon	£5.99	☐
GODS OF EDEN	Andrew Collins	£7.99	☐
ADVENTURES IN WONDERLAND	Sheryl Garratt	£7.99	☐
SBS	John Parker	£6.99	☐
THE COWBOY WAY	David McCumber	£6.99	☐
TILL THE SUN GROWS COLD	Maggie McCune	£6.99	☐
THE MAN WHO ATE EVERYTHING	Jeffrey Steingarten	£6.99	☐
CLOSE TO THE WIND	Pete Goss	£6.99	☐
ORDINARY DECENT CRIMINAL	Gerry Stembridge	£6.99	☐
ZARAFA	Michael Allin	£6.99	☐

Headline books are available at your local bookshop or newsagent. Alternatively, books can be ordered direct from the publisher. Just tick the titles you want and fill in the form below. Prices and availability subject to change without notice.

Buy four books from the selection above and get free postage and packaging and delivery within 48 hours. Just send a cheque or postal order made payable to Bookpoint Ltd to the value of the total cover price of the four books. Alternatively, if you wish to buy fewer than four books the following postage and packaging applies:

UK and BFPO £4.30 for one book; £6.30 for two books; £8.30 for three books.

Overseas and Eire: £4.80 for one book; £7.10 for 2 or 3 books (surface mail)

Please enclose a cheque or postal order made payable to *Bookpoint Limited*, and send to: Headline Publishing Ltd, 39 Milton Park, Abingdon, OXON OX14 4TD, UK.
Email Address: orders@bookpoint.co.uk

If you would prefer to pay by credit card, our call team would be delighted to take your order by telephone. Our direct line 01235 400 414 (lines open 9.00 am–6.00 pm Monday to Saturday 24 hour message answering service). Alternatively you can send a fax on 01235 400 454.

Name ....

Address ....

....

....

If you would prefer to pay by credit card, please complete:
Please debit my Visa/Access/Diner's Card/American Express (delete as applicable) card number:

Signature .... Expiry Date ....